Herbert Waldmann, Petra Janning
Chemical Biology
A Practical Course

Related Titles from Wiley-VCH

C. Schmuck, H. Wennemers (eds.)

Highlights in Bioorganic Chemistry

2004, ISBN 3-527-30656-0

H.-D. Höltje, W. Sippl, D. Rognan, G. Folkers

**Molecular Modeling –
Basic Principles and Applications (2nd Ed.)**

2003, ISBN 3-527-30589-0

T. K. Lindhorst

**Essentials of Carbohydrate Chemistry and Biochemistry
(2nd Ed.)**

2003, ISBN 3-527-30664-1

C. W. Sensen

Essentials of Genomics and Bioinformatics

2002, ISBN 3-527-30541-6

N. Sewald, H.-D. Jakubke

Peptides: Chemistry and Biology

2002, ISBN 3-527-30405-3

Herbert Waldmann, Petra Janning

Chemical Biology

A Practical Course

**WILEY-
VCH**

WILEY-VCH Verlag GmbH & Co. KGaA

Prof. Dr. Herbert Waldmann
Dr. Petra Janning

Max-Planck-Institute of Molecular Physiology
Otto-Hahn-Str. 11
44227 Dortmund
Germany
and
University of Dortmund
Organic Chemistry III
Otto-Hahn-Str. 6
44227 Dortmund
Germany
herbert.waldmann@mpi-dortmund.mpg.de
petra.janning@mpi-dortmund.mpg.de

■ This book was carefully produced. Nevertheless, authors and publisher do not warrant the information contained therein to be free of errors. Readers are advised to keep in mind that statements, data, illustrations, procedural details or other items may inadvertently be inaccurate.

Cover design: G. Schulte, Dortmund

Library of Congress Card No.: applied for
A catalogue record for this book is available from the British Library.

**Bibliographic information published
by Die Deutsche Bibliothek**
Die Deutsche Bibliothek lists this publication in the Deutsche Nationalbibliografie; detailed bibliographic data is available in the Internet at <http://dnb.ddb.de>.

© 2004 WILEY-VCH Verlag GmbH & Co. KGaA, Weinheim

Printed in the Federal Republic of Germany
Printed on acid-free paper

Typesetting   K+V Fotosatz GmbH, Beerfelden
Printing   Strauss GmbH, Mörlenbach
Bookbinding   Litges & Dopf Buchbinderei GmbH, Heppenheim

ISBN   3-527-30778-8

Preface

Chemistry and Biology are at the heart of the Natural Sciences (together with Physics of course) but during the last decades both sciences have undergone a very different development. Chemistry triumphantly emerged from the 19th century and kept celebrating similar triumphs for instance in the development of new materials (as one would say in these days) and treatments for diseases until well into the 20th century. However, gradually it became clear that many if not all of the molecular entities in the focus of more mainstream chemistry research can be made by means of the methodology available and the discovery of new "name reactions" has virtually come to a standstill in the 1990s. It would be wrong and testimony to a lack of insight to state that chemistry in these days is a mature science. A lot is left to be done in the development of chemistry methodology to a truly efficient and reliable science, and one should realize that the really challenging synthesis endeavors can only be mastered by a fairly small group of highly experienced teams operating at the forefront of research. But it appears to be clear as well that much of the intellectual adventure and challenge that characterized chemistry intrinsically has shifted to its neighboring discipline biology. Biology used to be a largely phenomenologically dominated science until well into the second half of the 20th century. However, since the 1960s gradually but irresistibly the molecular aspects of biology moved into the focus of research and today many elder biologists complain that the younger generation wants to juggle genes but does not know how to classify trees anymore. Indisputably biology has celebrated numerous successes in the last decades and at the latest with the successful sequencing of the human and other genomes it has developed into "big time" science. At the turn of the 20th to the 21st century biology is in a similar state of success as chemistry was a century before. It is paramount to note though that – similar to the relationship between physics and chemistry at the turn of the 19th to the 20th century – the objects, topics, methods and most importantly the dimensions of research in these neighboring disciplines converge. The targets of chemistry research in these days are no longer the small strained molecules of the 1950s and 1960s. The strength of the covalent bond and the verification or falsification of theories surrounding it as well as the unraveling of reaction mechanisms have lost much of their lure. Rather large naturally occurring molecules, supramolecular architectures, the "synthesis" of properties and the chemistry of (bio)polymers have moved into focus. The size of the objects of interest has gradually increased.

Biology has moved from the more descriptive, phenomenological level to a molecular science where the structure of and the interaction between (bio)macromolecules with each other, with low molecular weight compounds and with supramolecular structures like membranes and the cytoskeleton and their directing influence on biological phenomena are among the hottest topics of research. Where the borderlines of research begin to vanish between two neighboring disciplines new interfacial areas are being created. This was the case at the turn of the 19th to the 20th century when Physical Chemistry and Chemical Physics emerged. Fritz Haber held the chair for "Chemie der Gasphasenreaktionen" during the time he spent in Karlsruhe, i.e. when he developed the conversion of nitrogen and hydrogen into ammonia for which he received the Nobel Prize in Chemistry – formally there was no Physical Chemistry at the time.

At the turn of the 20th to the 21st century the relationship between Chemistry and Biology is similar, "Chemical Biology" being the name of the new discipline that is being shaped (see also below). Today Physical Chemistry is one of the three pillars of teaching and research in Chemistry Departments worldwide, and it is foreseeable that in the not-too-distant future a branch focusing on the borderline between Chemistry and Biology will be the fourth pillar (unexpectedly and curiously "Biochemistry" did not become it, primarily because it developed into a sub-field of Biology rather than Chemistry; see below).

If this will happen – and we do not doubt it – academic training in Chemistry and Biology will have to change and meet the demands and expectations of the new generations of students seeking the intellectual adventure. This will require new lecture courses which can be set up relatively easy. However, without additional practical training a truly interdisciplinary education that is the prerequisite for subsequent interdisciplinary research cannot be achieved.

To the best of our knowledge – and we will happily have us corrected for being unaware of similar attempts – the course summarized in this book is the first to be introduced into University teaching. It has successfully been completed by Chemistry students at the University of Dortmund and by Chemistry and Biology students participating in the graduate education offered by the International Max Planck Research School in Chemical Biology jointly run by the Chemistry and Biology Departments of the University of Dortmund and the University of Bochum together with the Max-Planck-Institute of Molecular Physiology, Dortmund.

It includes fundamental experiments like isolation of an enzyme, solid-phase peptide or oligonucleotide synthesis and proteomics-related experiments as well as experiments of more specialized nature that emerged from joint research projects between chemistry- and biology-oriented groups within the Max-Planck-Institute during the last decade. In the ideal case one would design two or more experiments directly related to each other to clearly point out the interdisciplinary nature of Chemical Biology research. For instance the synthesis of lipidated peptides and proteins, the determination of their biophysical properties and their application in Cell Biology experiments would be an example. But to date this can only be successfully realized in the context of a practical training course in very few examples.

We are aware of the facts that there is a certain bias towards experiments related to our own research and that several of the experiments described here require access to instrumentation that may not be routinely available in any Chemistry and Biology Department, but in order to set up a first practical course this proved to be necessary for feasibility reasons.

Each chapter is structured in the same way: after a short abstract and some information about the learning targets of the experiment a theoretical background is given. This is focused on the information needed to understand the experiment. It does not give a complete overview of a special topic, and wherever necessary references for further reading are given. This theoretical background is followed by the experimental procedures, which are divided into two parts. In the part "preparations" information about preparatory work to be carried out by the supervisors before the students start the experiments is given. In the second part the execution of the experiment during the practical course is described. Of course, it may be necessary to deviate from this routine due to very individual reasons (e.g. time schedule of the students) and in each environment it is necessary to optimize, which part of an experiment has to be carried out before the students start and which parts are carried out by the students during the practical course. At the end of each chapter references are given.

Typically, in the practical course groups of two or three students carried out one experiment per week (ca. five hours per day), and the entire course was completed within one semester for a group of up to 20 students. The experiment described in chapter 10 (proteomics) was divided into two weeks. In the first week the 2D-gel electrophoresis, in the second week the tryptic digest and the mass spectrometric identification of the proteins was carried out. Each group of students was supervised and examined by one or two experienced graduate students. This kind of training is fairly intensive and requires substantial input and manpower. However, we feel that it is worth the effort, and feedback by students who passed the course (drop-out rates were virtually zero) was very positive. We realize that the practical course described in this book is only a first attempt. It leaves room for substantial improvement and will certainly be surpassed in the future. We would like to invite all colleagues who are aware of illustrative and educative experiments that would meet the demands sketched above to approach us for inclusion of such experiments into future editions.

Finally we would like to express our gratitude to our coworkers and colleagues who contributed to adapting the described experiments to a format that can routinely and reliably be run in a student training course. Their names are mentioned with the description of the corresponding experiments. We are also grateful to Gudrun Walter and Frank Weinreich, Wiley-VCH for editorial help and encouragement.

Dortmund, January 2004

Petra Janning
Herbert Waldmann

Contents

Contents XI

Abbreviations

2-D	two dimensional
A	adenosine
AA	amino acid
abbrev.	abbreviation
Ac	acetate/acetyl
ACE	angiotensin-converting enzyme
AdoMet	S-adenosyl methionine
ADP	adenosine diphosphate
AIDS	acquired immune deficiency syndrome
approx.	approximately
APT1	acyl protein thioesterase 1
aq.	aqueous
ATP	adenosine triphosphate
B	base
bAP	biotinylated alkaline phosphatase
BG	biotin-galactose
Bhoc	benzhydrylcarbonyl
Boc	butyloxycarbonyl
BOP	(benzotriazol-1-yloxy)tris(dimethylamino)phosphoniumhexa-fluorophosphate
bp	base pairs
BPB	bromophenol blue
BSA	bovine serum albumin
Bu	butyl
iBu	isobutyryl
Bz	benzoyl
C	cytidine
ca.	circa
cf.	confer
CHAPS	3-[(3-Cholamidopropyl)dimethylammonio]-1-propanesulfonate
cmc	critical micelle concentration
CNE	cyanoethylester
CoA	coenzyme A

CPG	controlled pore glass
DABCYL	4-(4-dimethylaminophenylazo)benzoyl
DAG	diacylglycerol
DCC	dicyclohexylcarbodiimide
DCM	dichloromethane
ddH$_2$O	double distilled water
DDI	DNA-directed immobilization
DEAE	diethyl aminoethyl (coupled to sepharose beads)
DHB	2,5-dihydroxybenzoic acid
DIC	N,N-diisopropylcarbodiimide
DIPEA	N,N-diisopropylethylamine
DMAP	4-(dimethylamino)pyridine
DMF	dimethylformamide
DMSO	dimethylsulfoxide
Dmt	dimethoxytrityl
DNA	deoxyribonucleic acid
ds	double strand
DTE	dithioerythritol
DTT	dithiothreitol
e.g.	for example
EDTA	ethylene diamine tetraacetic acid
ELISA	enzyme-linked immunosorbent assay
eq.	equivalent
ER	endoplasmic reticulum
ESI	electrospray ionization
et al.	et altera
etc.	et cetera
Far	farnesyl
FITC	fluorescein 5-isothiocyanate
FKBP	FK506 binding protein
fl.	full length
Fmoc	9-fluorenylmethoxycarbonyl
FPLC	fast protein liquid chromatography
FPP	farnesyl pyrophosphate
FRET	fluorescence resonance energy transfer
FTase	farnesyl transferase
G	guanosine
Gal	galactose
GalNAc	N-acetylgalactosamine
GalNAz	N-azidoacetylgalactosamine
GAP	GTPase activating protein
GC	gas chromatography
GDP	guanosine diphosphate
GEF	guanine nucleotide exchange factor
Ger	geranyl

GFP	green fluorescent protein
Grb2	growth factor receptor bound protein 2
GTP	guanosine triphosphate
HATU	N-[dimethylamino)-1H-1,2,3-triazolo[4,5-b] pyridin-1-ylmethylene]-N-methylmethanaminium hexafluorophosphate N-oxide
HBSS	Hank's Balanced Salt Solution
HBTU	N-[(1H-benzotriazol-1-yl)(dimethylamino)methylene]-N-methyl-methanaminium hexafluoro-phosphate N-oxide
HEPES	N-(2-hydroxyethyl)piperazine-N'-(2-ethanesulfonic acid)
HIV	human immunodeficiency virus
HOBt	N-hydroxybenzotriazole
HPLC	high performance liquid chromatography
HRP	horseradish peroxidase
i.e.	id est
IC_{50}	inhibitor concentration 50
Ig	immune globulin
IP_3	myo-inosite-1,4,5-triphosphate
IPG	immobilized pH gradient
IR	infra red
LAH	lithium aluminium hydride
m/z	mass to charge ratio
MALDI	matrix-assisted laser desorption ionization
MALDI-TOF	matrix-assisted laser desorption ionisation – time of flight
MAP	mitogen-activated protein
Me	methyl
MeCN	acetonitrile
mod.	modified
MPI	Max Planck Institute
mRNA	messenger ribonucleic acid
MS	mass spectrometry
MTases	methyltransferases
NBD	7-nitro-1,2,3-benzoxadiazole
NF-AT	necrosis factor AT
NMR	nuclear magnetic resonance
no.	number
Nu	nucleophile
OD	optical density
P_i	inorganic phosphate
P.PH.	potato phosphorylase
PAGE	polyacrylamide gel electrophoresis
Pal	palmitoyl
PalCoA	palmitoyl co-enzyme A
PalTase	palmitoyl transferase
Pbf	2,2,4,6,7-pentamethyl-dihydrobenzofurane-5-sulfonyl-
PBS	phosphate-buffered saline

PC12 cells	phaeochromocytoma cells
pcCMT	prenylcysteine carboxyl methyltransferase
PCR	polymerase chain reaction
PenStreb	Penicillin/Streptomycin
PEP	Program for Engineering Peptides
PG	protection group
Pip	piperidine
PIP_2	phosphatidylinoside-4,5-bisphosphate
PNA	peptide nucleic acids
POPC	1-palmitoyl-2-oleylphosphatidylcholine or 1-hexadecanoyl-2-[*cis*-9-octadecenoyl]-*sn*-glycero-3-phosphocholine
PP	pyrophosphate
PP1	pyrazolo[3,5-d]pyrimidine-based inhibitor
PSD	post source decay
PyBOP	(benzotriazol-1-yloxy)tri(pyrrolidino)phosphonium
Pyr	pyridine
quant.	quantitative
QSAR	quantitative structure-activity relationship
Rab	Ras-like protein from rat brain
R	ratio
Ras	rat-adeno-sarcoma
R_{max}	ratio of maximum
R_{min}	ratio of minimum
Rib	ribose
RNA	ribonucleic acid
RP-HPLC	reversed phase high performance liquid chromatography
RP-LC-MS	reversed phase liquid chromatography mass spectrometry
rpm	revolutions per minute
rt	room temperature
S.Ph	sucrose phosphorylase
s/n	signal to noise
SAM	*S*-adenosyl-*L*-methionine
SAR	structure-activity relationship
SDS	sodium dodecyl sulfate
SDS-PAGE	sodium dodecyl sulfate polyacrylamide gel electrophoresis
SLBA	sugar-lectin binding assay
SNP	single nucleotide polymorphisms
Sos	1596-residue product of the *Son of Sevenless* gene
SPE	solid phase extraction
SPPS	solid phase peptide synthesis
ss	single strand
sSMCC	Sulfosuccinimidyl-4-(N-maleimidomethyl)cyclohexane-1-carboxylate
STV	streptavidin
STV-HRP	streptavidin-horseradish peroxidase

T	thymidine
TAMRA	6-carboxytetramethylrhodamine
TCA	trichloroacetic acid
TEA	triethylamine
TEMED	tetramethylethylenediamine
TFA	trifluoroacetic acid
THF	tetrahydrofuran
TIS	tri*iso*propyl silane
TOF	time of flight
Tris	tris-(hydroxymethyl)aminomethane (TRIZMA)
Tris-HCl	tris(hydroxymethyl)-aminomethane hydrochloride
UV	ultra-violet
v/v	volume per volume
vol	volume
w/v	weight per volume
wt	wild type

1
Introduction: Chemical Biology –
A New Science at the Crossroads of Chemistry and Biology

Rolf Breinbauer and Herbert Waldmann

During the last decade the life sciences have undergone dramatic changes. Biologists have deciphered the genetic codes of various organisms and determined the structure and intimate interplay of numerous macromolecules and small compounds. Chemistry has developed the methodology to synthesize – in principle – each biologically relevant small molecule as well as macromolecules such as proteins and nucleic acids. The development of new techniques and the application of methods used in physics have yielded a plethora of analytical methods with which it is possible to monitor biological and biochemical processes in precise detail in the test tube, in living cells and even in entire organisms. The insight gained by these and earlier investigations has provided unambiguous proof of the fact that molecular interactions and chemical transformations are at the heart of biology and that all biological phenomena which we can analyze today can ultimately be traced back to chemical processes: biology is molecular.

However, the various genomics and proteomics projects carried out worldwide are yielding a wealth of information, data bits provided in a language that we only partly understand. The task lying before us is to group together the letters making up the genome sequences into words, sentences, paragraphs and chapters, ultimately assembling the book of life. Taking a closer look at the prerequisites needed for the successful execution of this process reveals that simply knowing which genes are available and can potentially be expressed (genomics), is not sufficient. Also it will not suffice to determine which proteins are actually present in a living cell under any given conditions (proteomics). Ultimately the fate of a cell will be determined by the interactions of proteins with each other, with other biomacromolecules, with supramolecular structures like membranes and the cytoskeleton, and with small molecules binding to the gene products and modulating their activity (Fig. 1.1).

On the one hand this insight makes clear that chemistry may turn out to be the central science in the quest for understanding the molecular basis of life. This is so because the study of interactions between molecules, whether they are small or large, and the methodology to prepare them and to prepare new custom-made compounds are at the heart of chemistry and constitute the key expertise of chemists. On the other hand the undisputed powers of chemistry will need to be devoted to addressing the problems unraveled by research in the biological sciences.

Chemical Biology: A Practical Course.
Edited by Herbert Waldmann and Petra Janning
Copyright © 2004 WILEY-VCH Verlag GmbH & Co. KGaA, Weinheim
ISBN: 3-527-30778-8

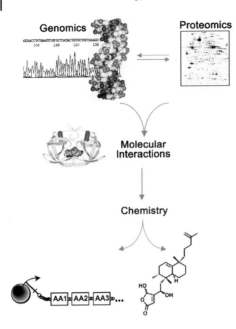

Fig. 1.1 The need for integration of Chemistry and Biology.

Between these two poles a new science has been and is still being created: Chemical Biology. Chemical biology may be defined as the development and use of chemistry techniques for the study of biological phenomena [1, 2]. It approaches biological problems from the point of view of the chemist and employs as one of its central methods the ability to design and prepare compounds with a pre-determined set of properties that can be subsequently applied to probe biological systems.

Currently major fields of interest in chemical biology are for instance, the study of biological signaling and transport or regulation of gene transcription. In chemical biology research a given biological system is typically perturbed by means of a chemical compound and then analyzed. Also, specifically labeled compounds may be introduced into cells and their fate is followed with appropriate biophysical methods. The methodology it employs differentiates chemical biology from current biochemistry and the earlier approaches commonly taken in bioorganic and biological chemistry. Biochemistry today mostly employs genetics and molecular biology methods as its core techniques. (This was not always the case however. In the 1950s and 1960s much of the research carried out by biochemists qualified for the definition of "Chemical Biology" as given above.) The term "Bioorganic Chemistry" describes research efforts focusing on an understanding and imitation of transformations carried out by living organisms by means of model compounds mimicking for instance the way in which enzymes catalyze chemical reactions. Biological chemists primarily focused their efforts on unraveling biosynthetic pathways, for instance by preparing isotope-labeled biosynthesis intermediates and analogs thereof [3, 4]. These definitions describing the different research ac-

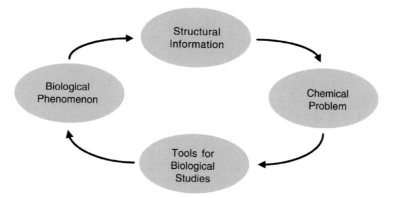

Fig. 1.2 Interplay between organic synthesis and biology in chemical biology research.

tivities at the crossroads of chemistry and biology since the 1950s should not be taken as strict definitions or as dogmas. Rather, the different fields overlap with each other and there is a gradual transition between them.1 Introduction: Chemical Biology – A New Science at the Crossroads of Chemistry and Biology

In general, chemical biology has gained rapidly increasing interest worldwide during the last decade, and today it is among the fastest growing areas of investigation in the Natural Sciences in general and in Chemistry in particular. For summaries of various research activities in this field the reader is referred to several reviews [1, 2, 5–17]. In order to illustrate the chemical-biological approach several case studies are presented below exemplifying the scope and the power of chemical biology.

Chemical biology usually starts by analyzing a biological system or phenomenon of interest (Fig. 1.2).

In this analysis structural information is deduced concerning the structure of biomacromolecules involved in a particular biological phenomenon for instance, or the structure of small molecules which interact with these macromolecules. This structural information is then employed to define unsolved chemical problems, i.e. the development of new methods for the synthesis of low molecular weight compounds such as natural products and analogs thereof, and of biomacromolecules. Also the design and synthesis of inhibitors that can be used to perturb and probe biological systems are of major interest. Once methods of accessing the desired compounds have been devised and developed, the newly prepared compounds are employed in appropriately designed biological and/or biochemical experiments. The results gleaned may then give rise to a better understanding of the biological problem. They may also highlight new structural features thus forming the basis for a new round of investigation.

1.1
Case Study 1: The Chemical Biology of Lipidated Proteins belonging to the Ras Superfamily

Lipid-modified proteins play important roles in numerous biological processes such as signal transduction, vesicular trafficking and organization of the cytoskeleton (Fig. 1.3). In order to study these processes in precise molecular detail methods for the synthesis of differently lipidated peptides and proteins were developed which may also carry further reporter groups and labels.

The Ras proteins are farnesylated and palmitoylated post-translationally and play a decisive role in signal transduction (Fig. 1.4). Mutations of Ras are found in approximately 30% of all human cancers making these proteins very important targets for biology and chemical biology. With the help of conjugates which link differently lipidated peptides to a truncated oncogenic Ras protein the examination of numerous differently lipid-modified proteins became possible.

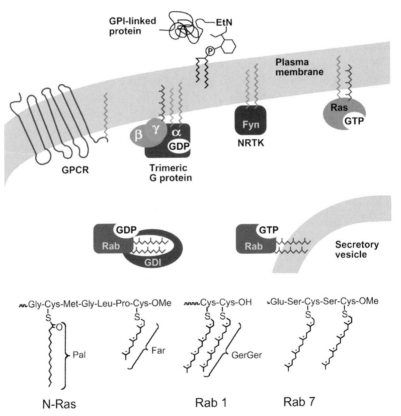

Fig. 1.3 Schematic representation of different lipid-modified proteins and representative lipidation pattern of Ras and Rab proteins. Pal, palmitoyl residue; Far, farnesyl residue; GerGer, geranylgeranyl residue.

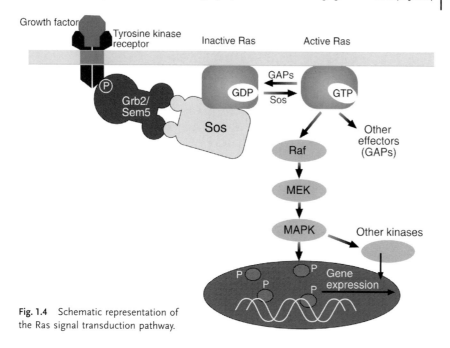

Fig. 1.4 Schematic representation of the Ras signal transduction pathway.

The study of correctly lipidated Ras proteins was hampered by the fact that these membrane-bound proteins could not be expressed by molecular biology techniques or were only expressed in very small amounts. This called for the development of methods for the synthesis of fully biologically-functional Ras proteins. The development of efficient synthetic access to these modified proteins would permit the modification of lipid moieties at will and the introduction of additional reporter groups by which the fate of the semi-synthetic proteins might be followed inside cells.

In a joint project, the groups of Kuhlmann, Wittinghofer, and Waldmann at the MPI Dortmund successfully combined organic synthesis, biophysics and cell biology techniques to yield new insight into a biological problem using a chemical biology approach [18, 19]. With regard to the protecting groups in particular, this approach required the development of a set of new techniques which allowed for selective deprotection under very mild conditions (i.e. pH 6–8, room temperature) since the farnesyl groups are acid sensitive and the palmitic acid thioesters are very base labile. The use of biocatalysis and noble metal catalysis proved to be key techniques in achieving this goal, resulting e.g. in the introduction of a set of new enzyme-labile urethane and ester protecting groups (Scheme 1.1).

Furthermore, a generally applicable method for the solid-phase synthesis of lipidated and additionally labeled peptides was developed. In this work knowledge gleaned from the development of new linker groups for solid-phase synthesis was successfully transferred. Thus, application of the hydrazide linker – originally de-

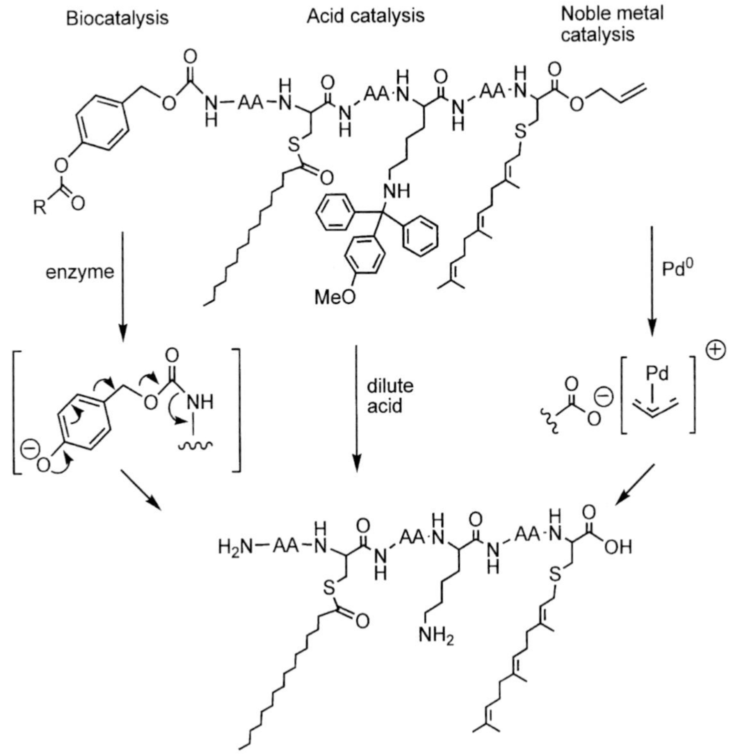

Scheme 1.1 Strategies for protecting groups during the synthesis of lipidated peptides.

veloped for traceless combinatorial synthesis – to solve the problems posed by lipopeptide chemistry allowed for the efficient synthesis of Ras and Rab peptides (Scheme 1.2) [20].

By means of these techniques peptides embodying the natural lipid parts of the Ras proteins, non-natural analogs thereof as well as fluorescence-labeled prenylated and/or palmitoylated Ras peptides were equipped with a reactive linker group such as the maleimido group, and coupled to C-terminally truncated human Ras proteins. The Ras hybrid proteins were employed as tools for membrane binding experiments e.g. via surface plasmon resonance and fluorescence-based assays employing vesicles as membrane models. Microinjection into PC12 cells revealed that the biological activity of Ras is not dependent upon a farnesylated residue and identified palmitoylation as a key regulatory step in the selective targeting of Ras to the plasma membrane (Fig. 1.5).

Thus, the hybrid proteins serve as suitable tools for biochemical, biophysical and biological experiments. These investigations resulted in the development of a new readout system that allows structural information to be correlated with the biological activity at the plasma membrane [21, 22].

Scheme 1.2 Generally applicable solid-phase approach for the rapid synthesis of lipidated peptides.

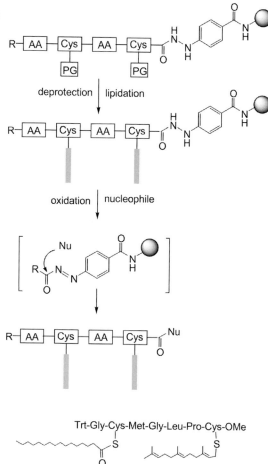

Trt-Gly-Cys-Met-Gly-Leu-Pro-Cys-OMe

C-terminus of the N-Ras protein
68% overall yield

1.2
Case Study 2: Chemical Biology of Rab Proteins

The Rab proteins are a sub-family of the ras-like GTPase superfamily with more than 60 members. Rab GTPases control a broad array of membrane docking and fusion events operating in processes ranging from fertilization to synaptic transmission (Fig. 1.6).

In order to function, Rab proteins require double modification with geranylgeranyl isoprenoids, which allows them to reversibly associate with membranes in a tightly controlled fashion. In order to study the biochemical properties and biolog-

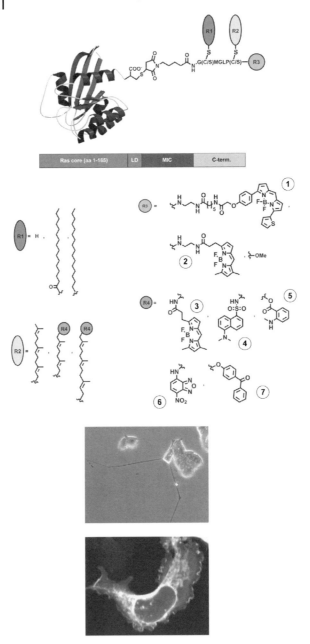

Fig. 1.5 Various synthetic Ras-hybrid proteins. These modified proteins were used as tools for membrane binding experiments. When microinjected into neuronal precursor cells, correctly lipidated oncogenic Ras proteins (either by means of organic synthesis or by the intracellular biosynthesis machinery) are localized to the plasma membrane and induce neurite outgrowth. The cellular location of the protein shown in the lower part of the figure is made visible by the incorporation of fluorescent labels.

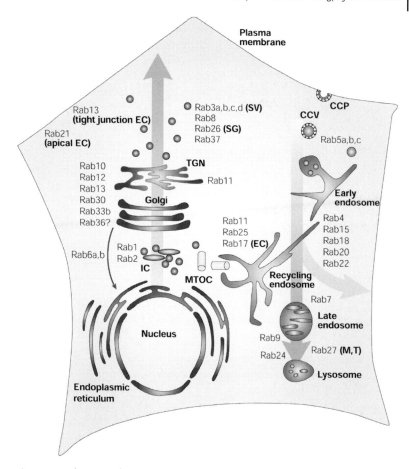

Fig. 1.6 Involvement of Rab proteins in vesicular transport processes (according to [23], reprinted with permission from *Nature Reviews Molecular Cell Biology*, copyright 2001, Macmillan Magazines Ltd.).

ical functions of the Rab proteins and for the study of protein complexes in which Rab proteins participate, methods were developed that provide prenylated Rab proteins with new functionalities such as fluorescence, photoreactivity or isoprenoid groups at non-native positions. The recently developed *in vitro* protein ligation method provided the necessary platform for combining large recombinant protein scaffolds with peptides generated by organic synthesis [24].

As a result of close cooperation between the groups of Alexandrov, Goody, and Waldmann at the MPI Dortmund, Rab-peptides were synthesized by means of the preparative methods developed in the context of the Ras projects detailed above. These peptides were equipped with an N-terminal cysteine and then coupled to a truncated Rab protein by means of the "expressed protein ligation" technique which gives access to Rab proteins activated as C-terminal thioesters (Fig. 1.7) [25].

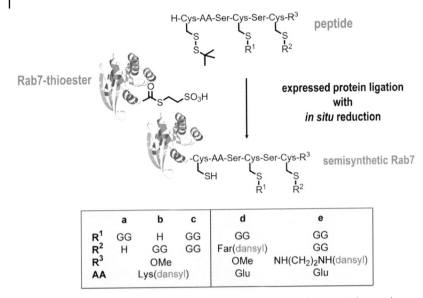

Fig. 1.7 Synthesis of Rab proteins by means of expressed protein ligation. Rab peptides containing different modifications were ligated to truncated Rab proteins.

	a	b	c	d	e
R^1	GG	H	GG	GG	GG
R^2	H	GG	GG	Far(dansyl)	GG
R^3		OMe		OMe	$NH(CH_2)_2NH$(dansyl)
AA		Lys(dansyl)		Glu	Glu

Fig. 1.7 Synthesis of Rab proteins by means of expressed protein ligation. Rab peptides containing different modifications were ligated to truncated Rab proteins.

These proteins were then employed on the one hand in a series of biochemical experiments providing precise data regarding the orchestration of events leading to sequential Rab prenylation. On the other hand the availability of Rab proteins carrying either one or two geranylgeranyl units opened up the opportunity to analyze the structure of the Rab proteins in complex with different chaperones [26] which determine their delivery to and extraction from different intracellular vesicles in the course of steering vesicular membrane fusion processes (Fig. 1.8).

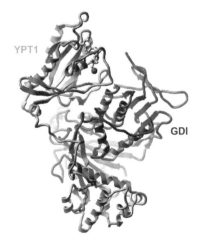

Fig. 1.8 Structure of semisynthetic Ypt1, a Rab homolog from yeast, in complex with the Rab guanine nucleotide dissociation inhibitor (GDI).

On the basis of the X-ray structures a mechanism was deduced to explain the selective extraction of Rabs from vesicular membranes in molecular detail. Notably, the structure determination of the Rabs in complex with their natural partner proteins was one of the long-standing problems in structural biology that could not be solved by several groups worldwide for more than a decade. This is a very illustrative example and clear-cut proof-of-principle for the impact that the combination of chemical and biological expertise can have.

1.3
Case Study 3: Identifying the Natural Biological Target of FK506 and the Design of Protein Dimerizers as a Research Tool in Chemical Biology

FK506 is a natural product which was isolated from fungi by Japanese researchers in the 1980s. In initial screens it exhibited interesting biological activity indicating that it may have potential as an immunosuppressant drug in organ transplantation (Scheme 1.3). This inspired great research activity on both the chemical synthesis of this natural product and the identification of the biological target with which this small molecule interacts [5]. Schreiber's group succeeded in the total synthesis of this complex polyketide product. During the course of this endeavor they gained a wealth of information about the intrinsic reactivity of this natural product, which they were able to exploit in the preparation of an affinity column of this molecule prepared by covalently attaching FK506 to a solid phase material via the exocyclic double bond. Using this chemical probe they identified a new protein FKBP ("FK506 binding protein") which binds to the eastern half of FK506 embodying the piperidincarboxylic acid structure. This moiety resembles the eastern half of the immunosuppressant rapamycin, which indeed also bound FKBP.

Although both FK506 and rapamycin bind FKPB and inhibit its intrinsic rotamase activity (that means, it catalyzes the cis/trans isomerization of amide bonds), it was shown that rapamycin inhibits a signal pathway other than that of FK506. Building

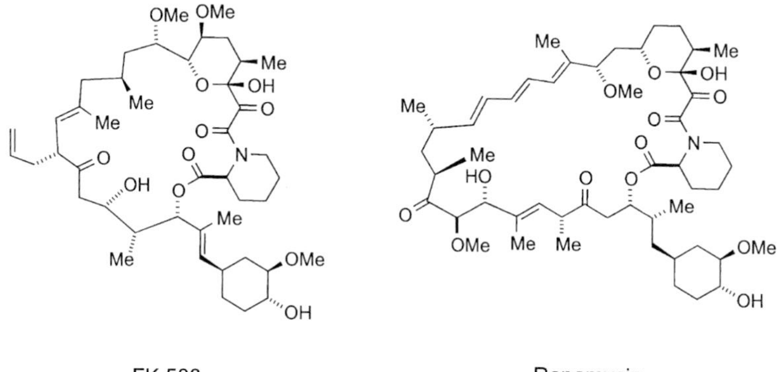

FK 506

Rapamycin

Scheme 1.3   Immunosuppressive natural products FK506 and rapamycin.

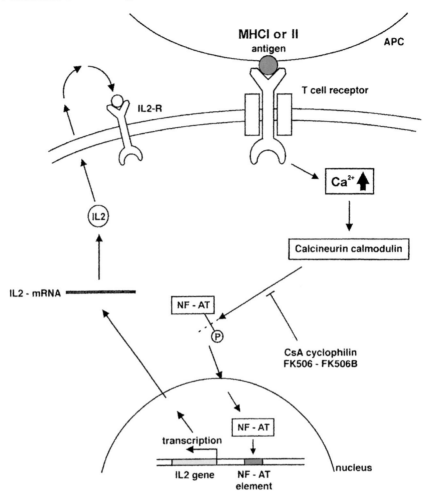

Fig. 1.9 Model of the function of calcineurin in T lymphocytes. Antigenic peptides are presented to the T lymphocytes by an antigen-presenting cell (APC) within a cell–cell interaction. Antigen binding activates the T cell receptor which initiates a signal chain leading to an increase in cytosolic Ca^{2+} and activation of calcineurin. The activated calcineurin cleaves an inhibitory phosphate residue from the transcription factor NF-AT. Complexes of the immunosuppressant FK506 with its binding protein inhibit calcineurin and disrupt the signal transmission to NF-AT [27]. (Courtesy of Wiley-VCH).

on the hypothesis that the ligand–receptor complex influences a molecule of another signal transduction cascade, the cellular binding partner of the "activated complex" of FK506 and FKBP was found, again with the help of affinity chromatography. This turned out to be the calmodulin-dependent phosphatase calcineurin; its enzymatic activity is inhibited by binding to FK506/FKBP, but is not influenced by FK506 or rapamycin. This enzyme inhibition hinders phosphorylation of the transcription factor NF-AT leading to the arrest of the immune system-relevant T cell cycle and thus to the observed immunosuppression (Fig. 1.9).

Having identified the natural binding partner of FK506, the groups of Schreiber and Crabtree designed chemical probes to initiate gene expression [28]. Using this approach, dimers of protein-binding molecules such as FK506, rapamycin or cyclosporin act as molecular glue to bring proteins together. For this it is necessary to synthesize hybrids between the proteins of interest and the respective protein-binding domains of the corresponding small molecules. Transcription factors are made up of functionally different domains which mediate binding to DNA and activation of transcription independent of one another. Since no covalent bonds are needed here it is possible to bring such domains close to each other with the help of dimeric ligands and thus to induce transcription. For this purpose, chimeric proteins were used, namely a DNA-binding protein linked to FKBP and an activation domain also linked to FKBP. These were expressed in cells. Treatment of these cells with the chemically-synthesized dimeric ligand FK1012 led to the activation of transcription, which could be stopped by addition of the monomer FK506 (Fig. 1.10). This concept of dimerizers turned out to be quite general and the Schreiber group has since applied this approach in many other ways [14].

1.4
Case Study 4: Covalent Trapping of Protein–DNA Complexes

Although the proposal of the double-helix structure by Watson and Crick can be regarded as the beginning of modern molecular biology, it should not be forgotten that if DNA always followed this canonical double-helix structure it would be devoid of any interesting biological activity. For its biological function (replication, transcription, repair, epigenetic modification, etc.) interaction with proteins is required and this is often accompanied by major structural changes in the DNA molecule [30]. In order to understand the molecular mechanisms of protein/DNA interaction it would be desirable to obtain structural information regarding not only the individual partners but both "actors" in the same scene, ideally "caught in the act". A means for achieving this is "mechanism-based trapping", which relies on a detailed picture of the enzyme mechanism to subvert the normal course of the reaction through chemical manipulation, thereby resulting in the accumulation of an ordinarily transient intermediate or a close analog.

An elegant example of such a trapping experiment has been provided by the group of Verdine [29]. Many organisms (especially bacteria) increase the information content of their genome by expanding the genetic alphabet from four (A, C, G, and T) to five or even more bases. The non-canonical bases are incorporated into the genome not by replication, but by covalent modification of canonical bases in DNA. One mode of such epigenetic covalent modification is the enzyme-catalyzed transfer of a methyl group from the cofactor S-adenosyl methionine (AdoMet) to cytosine producing 5-methyl-cytosine. In mammals, 5-methyl-C plays a key role in the epigenetic silencing of gene expression and consequently is in-

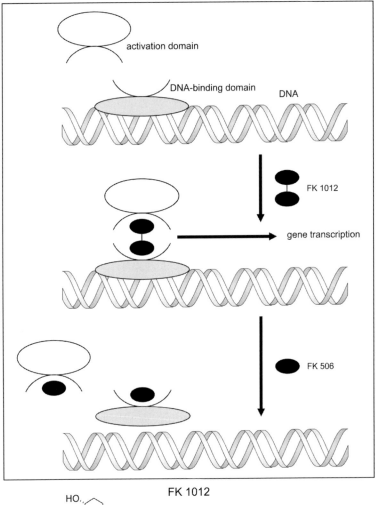

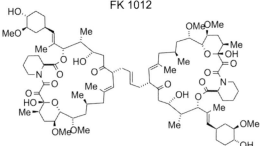

Fig. 1.10 Transcription activated by the addition of protein domain dimerizers.

dispensable for genomic stability, embryonic development, genomic imprinting, and X-chromosome inactivation. A plausible mechanism was proposed by which DNA cytosine-5 methyltransferases (MTases) employ covalent catalysis involving a conserved Cys residue on the enzyme to activate C5 for methylation and proton abstraction (Scheme 1.4, top row). A compelling piece of evidence in favor of the covalent catalysis scheme was the observation that DNA containing 5-fluoro-C, a mechanism-based inhibitor, became irreversibly attached to the MTase in the presence of AdoMet (Scheme 1.4, bottom row).

This chemically prepared probe not only allowed this mechanism to be studied by functional inhibition, but also enabled the crystallization of the covalent adduct formed between the derivatized DNA piece and the methyltransferase thus leading to the elucidation of its structure [30, 31]. The most dramatic feature of these MTase structures is the extrusion of the substrate cytosine from the DNA helix and its insertion into a deep concave pocket in the enzyme ("base-flipping"-mechanism; Fig. 1.11) [29].

1.5
Case Study 5: Cellular Imaging and *In Vivo* Labeling of Proteins with Fluorescent Probes

Understanding how chemical events within a cell are integrated requires real-time observation of molecules inside cells. The method of detection must sense both localization and chemical or physical changes of the molecules involved [9, 32, 33]. Because of its flexibility in implementation, non-invasiveness, and high sensitivity, fluorescence has emerged as the key tool for probing cellular biochemistry. A key requirement is the ability to attach fluorescent labels to the object of interest. This can be achieved by (i) using fluorescent indicators for ions (e.g. fluorescent Ca^{2+} sensors which allow the change in Ca^{2+}-ion concentration during cell signaling events to be followed visually) [34], (ii) preparing chimeras of protein with fluorescing proteins (e.g. conjugates with green fluorescent protein, which consist of 238 amino acids) [35, 36] or (iii) chemically attaching a small molecule fluorophore to a protein and microinjecting it into a cell (e.g. Section 1.1). As small molecule fluorophores exhibit certain advantages over GFPs, it is desirable to conceive a strategy in which small molecule fluorophores are attached to a protein of interest and in which the location of labeling can be genetically controlled. Tsien's group has recently introduced such a method for site-specific fluorescent labeling of recombinant proteins in living cells [36, 37]. The sequence Cys-Cys-Xaa-Xaa-Cys-Cys, where Xaa is a non-cysteine amino acid, is genetically fused or inserted within the protein, where it can be specifically recognized by a membrane-permeable fluorescein derivative with two As(III) substituents, FLASH, which fluoresces only after the arsenics bind to the cysteine thiols (Scheme 1.5) [38].

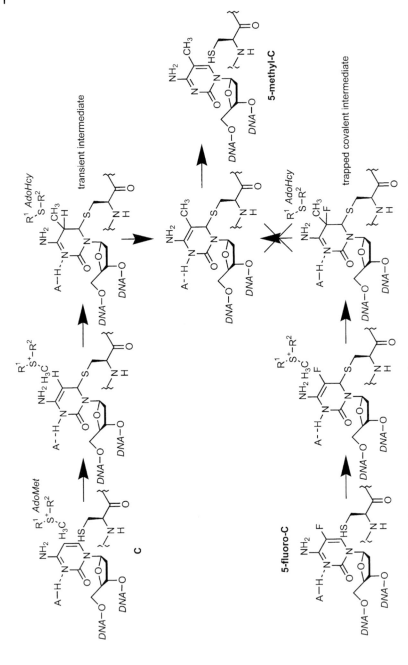

Scheme 1.4 Catalytic mechanism of cytosine-5-methyltransferase and mode of inhibition by 5-fluoro C in DNA [29].

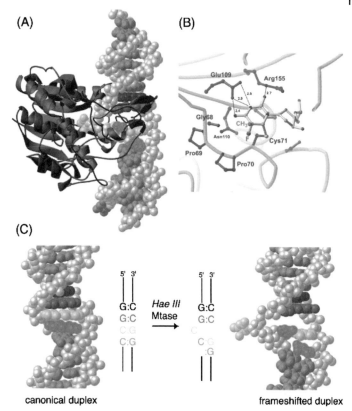

Fig. 1.11 Structure of the Hae II MTase trapped on an oligoncucleotide containing 5-fluoro-C. The protein is depicted as a ribbon trace, and the DNA as a space-filling model. The substrate cytosine is extra-helical and is inserted into the active pocket of the protein [29]. (Reprinted, with permission, from *Annual Review of Biochemistry*, Volume 72 ©2003 by Annual Reviews *www.annualreviews.org*).

1.6
Case Study 6: Modulating Cell Surface Architecture using Chemical Tools

The interactions that take place at the surface of a cell are of critical importance to the cell cycle and to the communication of cells within complex tissues. Molecular alteration of the cell surface would therefore change the presentation of the cell to the outside world and affect processes including cell–cell adhesion and virus–cell interactions. Thus there is tremendous interest in chemical methods that modulate cell-surface molecules with a view to investigating their function in the context of intercellular communication [9].

A very interesting chemical reaction for chemical cell-surface remodeling is the Staudinger ligation which has been developed by the group of Bertozzi [39, 40]. This reaction occurs between two abiotic functional groups, a specifically derivatized phosphine and an azide, to produce an amide-linked adduct.

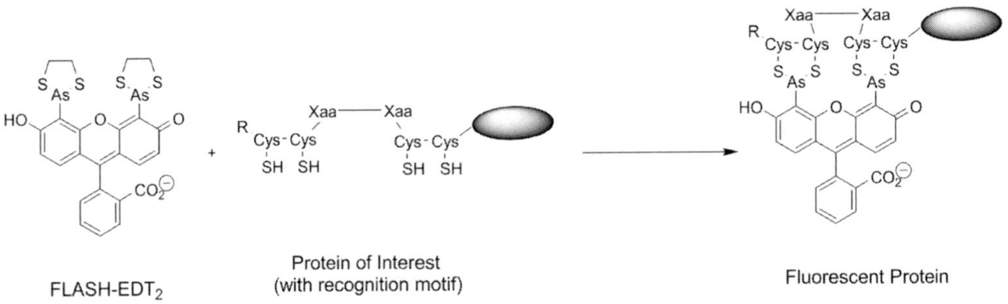

FLASH-EDT₂

Protein of Interest
(with recognition motif)

Fluorescent Protein

Scheme 1.5 Membrane-permeable dye FLASH, fluoresces after binding to proteins containing a Cys-Cys-Xaa-Xaa-Cys-Cys motif.

Post-translational modification of proteins is recognized as a major point of diversification that distinguishes the proteomes of higher organisms from simple organisms. Glycosylation is the most complex (and hence most difficult to study) form of post-translational modification and is known to regulate many aspects of protein function. The two major types of glyclosylation are N-linked (attached to Asn residues) and O-linked (attached to Ser or Thr residues). The predominant form of O-linked glyocosylation is the mucin-type, characterized by an initial N-acetylgalactosamine (GalNAc) residue a-linked to the hydroxyl groups of the Thr or Ser side-chains. Bertozzi et al. have reported a strategy for labeling such mucin-type O-linked glycoproteins with the bioorthogonal azide tag which exploits their one commonality: the conserved GalNAc residues [41]. The strategy involves feed-

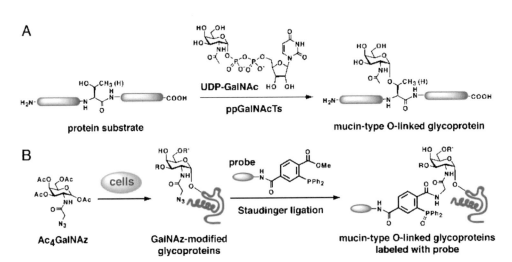

A

protein substrate

UDP-GalNAc

ppGalNAcTs

mucin-type O-linked glycoprotein

B

Ac₄GalNAz

cells

GalNAz-modified glycoproteins

probe

Staudinger ligation

mucin-type O-linked glycoproteins labeled with probe

Scheme 1.6 Strategy for metabolic labeling of mucin-type-O-linked glycoproteins with an azido GalNAc analog (BalNAz) for proteomic analysis with phosphine probes, exploiting the chemoselective Staudinger ligation reaction between azides and phosphines. R and R' are oligosaccharide elaborations form the core a-GalNAc residue (Reprinted, with permission, from Proc. Nat. Acad. Sci. USA, Volume 100 © 2003 by National Academy of Sciences, USA [41]).

ing cells an azido GalNac analog (N-azidoacetylgalactosamine, GalNAz) that is metabolically incorporated into mucin-type O-linked glycoproteins. Glycoproteins expressed at endogenous levels in mammalian cells were labeled with GalNAz, enabling their distinction from complex cell lysates with phosphine tags, which can be linked to fluorescence probes (Scheme 1.6).

1.7
Case Study 7: Allele-specific Inhibition of Kinases

For more than a century chemical compounds have been used for perturbing biological systems and looking for a change of phenotype [7, 14–17]. Several compounds have been identified which turned out to be very useful for biological investigations, e.g. the natural products such as brefeldin (cellular vesicle transport), okadaic acid (study of phosphatases) or colchicine (cell cycle arrest). The central problem of such an approach is the identification of small molecules which interact specifically with a desired protein target. This task becomes increasingly difficult with increasing numbers of related proteins of a certain class in the system under investigation, e.g. there are several hundred protein kinases in a human cell, all playing significant biological roles. Shokat *et al.* have recently presented an ingenious solution to this problem by combining mutation genetics and rationally designed inhibitors [42, 43]. Protein kinases are a family of enzymes which use ATP to attach phosphate groups to other proteins or small molecule substrates. The active site of protein kinases is well conserved and makes specific inhibition of a desired kinase challenging. Indeed there are general inhibitors which address the active sites of this enzymes promiscuously and inhibit all kinases, e.g. the pyrazolo[3,5-d]pyrimidine-based inhibitor PP1 (Fig. 1.12 a). If this inhibitor is modified with a more bulky group, such as a naphthyl group, it no longer fits into the active site of the wild-type enzymes (Fig. 1.12 b). On the other hand by applying standard techniques of molecular biology it is possible to engineer the active site of the kinase under investigation in such a way that the bulkier inhibitor fits into the active site of the mutant kinase (Fig. 1.12 d). Importantly, in the absence of inhibitor this mutant kinase remains completely biologically functional (Fig. 1.12 c). The Shokat group has applied this strategy to reveal the function of several protein kinases in the complex context of a cellular network. In one example they have prepared a mutant of the cyclin-dependent kinase Pho85 in *S. cerevisiae*, in which they exchanged the bulky amino acid Phe82 for a small glycine, thereby creating a hole in the active site. When a bulky inhibitor was added to a cell expressing this F82G mutant, gene expression analysis via DNA chip revealed that 853 genes exhibited a more than two-fold change in expression, thereby identifying those elements of the cellular network which are connected to the Pho85 kinase [44].

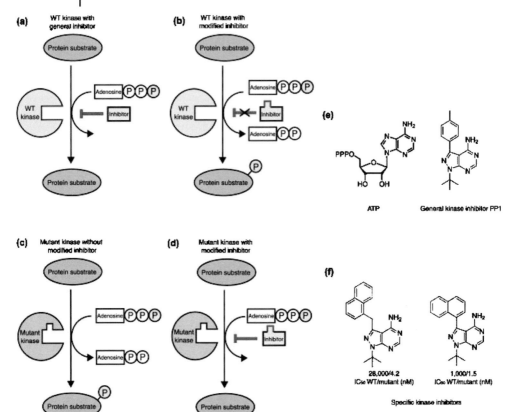

Fig. 1.12 Kinase-specific inhibition can be achieved by coupling chemistry and genetics (Reprinted, with permission, from *Annual Review of Cell and Developmental Biology, Volume 17 ©2001 by Annual Reviews www.annualreviews.org* [42]).

1.8
Conclusions

It will not have escaped the attention of the reader that the case studies described above were not exclusively single-group endeavors but involved many collaborating groups or scientists with different scientific backgrounds. It is important to realize that the key to trustworthy and efficient cooperation is that the partners share overlapping expertise and speak the same scientific language. Ideally, they should also be located in close spatial proximity which is paramount for cross-disciplinary communication. Overlapping expertise and common language, however, require that the scientists interacting with each other share a common training both in theory and in practice. It is the purpose of the practical course described in this book to provide a basis for training chemists and biologists in selected techniques and methods in this interfacial and rapidly developing area of science.

1.9
Bibliography

1 SCHREIBER, S.L. *Chem. Eng. News* 1992, October 26, 22–32.
2 FAMULOK, M., WALDMANN, H. *ChemBioChem* 2001, *2*, 3–6.
3 DUGAS, H. *Bioorganic Chemistry*, 3rd edn, Springer, New York, 1996.
4 FRUTON, J.S. *Proteins, Enzymes, Genes: The Interplay of Chemistry and Biology*, Yale University Press, New Haven and London, 1999.
5 ROSEN, M.K., SCHREIBER, S.L. *Angew. Chem. Int. Ed.* 1992, *31*, 384.
6 HINTERDING, K., ALONSO-DÍAZ, D., WALDMANN, H. *Angew. Chem. Int. Ed. Engl.* 1998, *37*, 688–749.
7 SCHREIBER, S.L. *Science* 2000, *287*, 1964–1969.
8 BREINBAUER, R., VETTER, I.R., WALDMANN, H. *Angew. Chem. Int. Ed.* 2002, *41*, 2878–2890.
9 COOK, B.N., BERTOZZI, C.R. *Bioorg. Med. Chem.* 2002, *10*, 829–840.
10 SHOKAT, K., VELLECA, M. *Drug Discovery Today* 2002, *7*, 872–879.
11 PETERSON, J.R., MITCHISON, T.J. *Chem. Biol.* 2002, *9*, 1275–1285.
12 FAMULOK, M., MAYER, G., BLIND, M. *Acc. Chem. Res.* 2000, *33*, 591–599.
13 WALDMANN, H. *Bioorg. Med. Chem.* 2003, *11*, 3045–3051.
14 SCHREIBER, S.L. *Bioorg. Med. Chem.* 1998, *6*, 1127–1152.
15 STOCKWELL, B.R. *TIBTECH* 2000, *18*, 449–454.
16 CREWS, C.M., SPLITTGERBER, U. *TIBS* 1999, *24*, 317–320.
17 ZHENG, X.F.S., CHAN, T.-F. *Drug Discovery Today* 2002, *7*, 197–205.
18 KADEREIT, D., KUHLMANN, J., WALDMANN, H. *ChemBioChem* 2000, *1*, 144–169.
19 KOCH, M., BREINBAUER, R., WALDMANN, H. *Biol. Chem.* 2003, *384*, 1265–1272.
20 LUDOLPH, B., EISELE, F., WALDMANN, H. *J. Am. Chem. Soc.* 2002, *124*, 5954–5955.
21 BADER, B., KUHN, K., OWEN, D.J., WALDMANN, H., WITTINGHOFER, A., KUHLMANN, J. *Nature* 2000, *403*, 223–226.

22 KUHN, K., OWEN, D.J., BADER, B., WITTINGHOFER, A., KUHLMANN, J., WALDMANN, H. *J. Am. Chem. Soc.* 2001, *123*, 1023–1035.
23 ZERIAL, M., MCBRIDE, H. *Nat. Rev. Mol. Cell Biol.* 2001, *2*, 107–117
24 EVANS JR., T.C., XU, M.-Q. *Chem. Rev.* 2002, *102*, 4869–4883.
25 ALEXANDROV, K., HEINEMANN, I., GOODY, R.S., DUREK, T., WALDMANN, H. *J. Am. Chem. Soc.* 2002, *124*, 5648–5649.
26 RAK, A., PYLYPENKO, O., DUREK, T., WATZKE, A., KUSHNIR, A., BRUNSVELD, L., WALDMANN, H., GOODY, R.S., ALEXANDROV, K. *Science* 2003, *302*, 646–650.
27 KRAUSS, G. *Biochemistry of Signal Transduction and Regulation*, 3rd edn, Wiley-VCH, Weinheim, 2003.
28 CRABTREE, G.R., SCHREIBER, S.L. *TIBS* 1996, *21*, 418.
29 VERDINE, G.L., NORMAN, D.P.G. *Annu. Rev. Biochem.* 2003, *72*, 337–366.
30 KLIMASAUSKAS, S., KUMAR, S., ROBERTS, R.J., CHENG, X. *Cell* 1994, *76*, 357–369.
31 REINISCH, K.M., CHEN, L., VERDINE, G.L., LIPSCOMB, W.N. *Cell* 1995, *82*, 142–153.
32 TSIEN, R.Y. *Nature Rev. Mol. Cell. Biol.* 2003, *4*, SS16–SS21.
33 ZHANG, J., CAMPBELL, R.E., TING, A.Y., TSIEN, R.Y. *Nature Rev. Mol. Cell. Biol.* 2002, *3*, 906–918.
34 TSIEN, R.Y., POZZAN, T. *Methods Enzymol.* 1989, *172*, 230–244.
35 TSIEN, R.Y. *Annu. Rev. Biochem.* 1998, *67*, 509–544.
36 ZIMMER, M. *Chem. Rev.* 2002, *102*, 759–781.
37 GRIFFIN, B.A., ADAMS, S.R., TSIEN, R.Y. *Science* 1998, *281*, 269–272.
38 ADAMS, S.R., CAMPBELL, R.E., GROSS, L.A., MARTIN, B.R., WALKUP, G.K., YAO, Y., LLOPIS, J., TSIEN, R.Y. *J. Am. Chem. Soc.* 2002, *124*, 6063–6076.
39 SAXON, E., BERTOZZI, C.R. *Science* 2000, *287*, 2007–2010.
40 DUBE, D.H., BERTOZZI, C.R. *Curr. Opin. Chem. Biol.* 2003, *7*, 616–625.

41 HANG, H.C., YU, C., KATO, D.L., BERTOZ-
zi, C.R. *Proc. Natl Acad. Sci. USA* **2003**,
100, 14846–14851.

42 SHOGREN-KNAAK, M.A., ALAIMO, P.J.,
SHOKAT, K.M. *Annu. Rev. Cell Dev. Biol.*
2001, *17*, 405–43.

43 ALAIMO, P.J., SHOGREN-KNAAK, M.A.,
SHOKAT, K.M. *Curr. Opin. Chem. Biol.*
2001, *5*, 360–367.

44 CARROLL, A.S., BISHOP, A.C., DERISI,
J.L., SHOKAT, K.M., O'SHEA, E.K. *Proc.
Natl Acad. Sci. USA* **2001**, *98* ,12578–
12583.

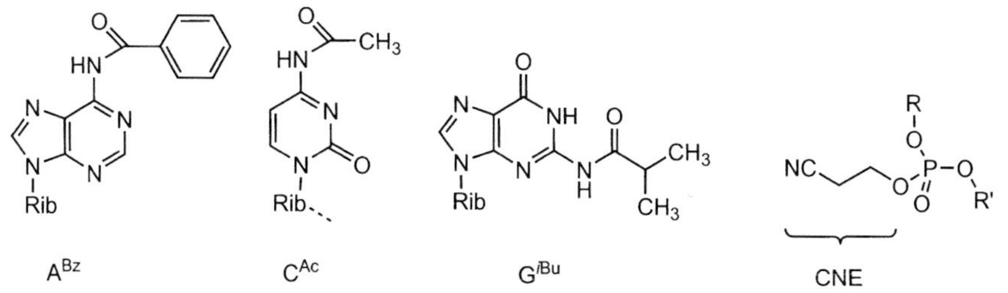

Scheme 2.3 Permanent protecting groups used in the synthesis of oligonucleotides.

base-protecting groups. In most cases, cyanoethylester is the protecting group of choice and can be cleaved via a β-elimination. These permanent protecting groups are removed upon treatment with a methanolic solution of ammonia within a reasonable period of time.

Protecting groups are called *temporary* if they have to be cleaved off several times during the synthesis. For example, when synthesizing a dinucleotide in solution, either the 5'-OH or the 3'-OH protecting group needs to be removed. During synthesis of large oligonucleotides these steps have to be repeated several times.

In solid phase synthesis an insoluble material acts as the protecting group for the 3'-hydroxy group and is removed by cleavage along with the protecting groups of the nucleobases at the end of the process. The 5'-hydroxy group however, requires only temporary protection. When the permanent protection groups are base labile, the temporary protecting groups should be acid labile. However, to prevent depurination (Scheme 2.2) only very mild conditions should be used. An optimal choice is a trityl group, especially the dimethoxytrityl group (Scheme 2.4). This group can be cleaved with 3% trichloroacetic acid in DCM in a few minutes. The colored trityl cation which is formed, allows the reaction on solid support to be controlled spectrophotometrically.

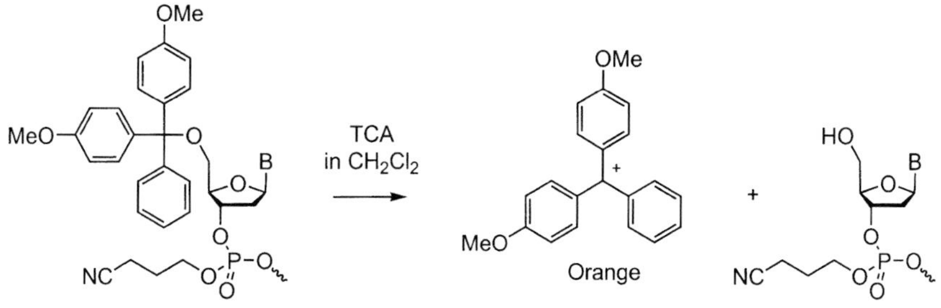

Scheme 2.4 Cleavage of the dimethoxytrityl protecting group (temporary protecting group).

2.3.2.3 General Strategy for DNA Synthesis

In contrast to biosynthesis, chemical synthesis is carried out in the 3′ to the 5′ direction. For solid phase synthesis, the monomer is bound at its 3′ end to the solid support. The most widely used method is the Caruthers method. The most commonly used solid support is controlled pore glass (CPG) which is chemically inert and incapable of swelling. It is functionalized with reactive amino groups, normally long-chain alkyl amino groups. It is necessary to have a connection between the nucleoside of the 3′ end and the solid support which allows cleavage of the oligonucleotide at the desired length. Normally, a succinyl linker is used. The synthesis (Scheme 2.5) begins with the cleavage of the 5′-hydroxy protecting group from the first nucleoside **1** bound to the solid support. In the next step, a build-

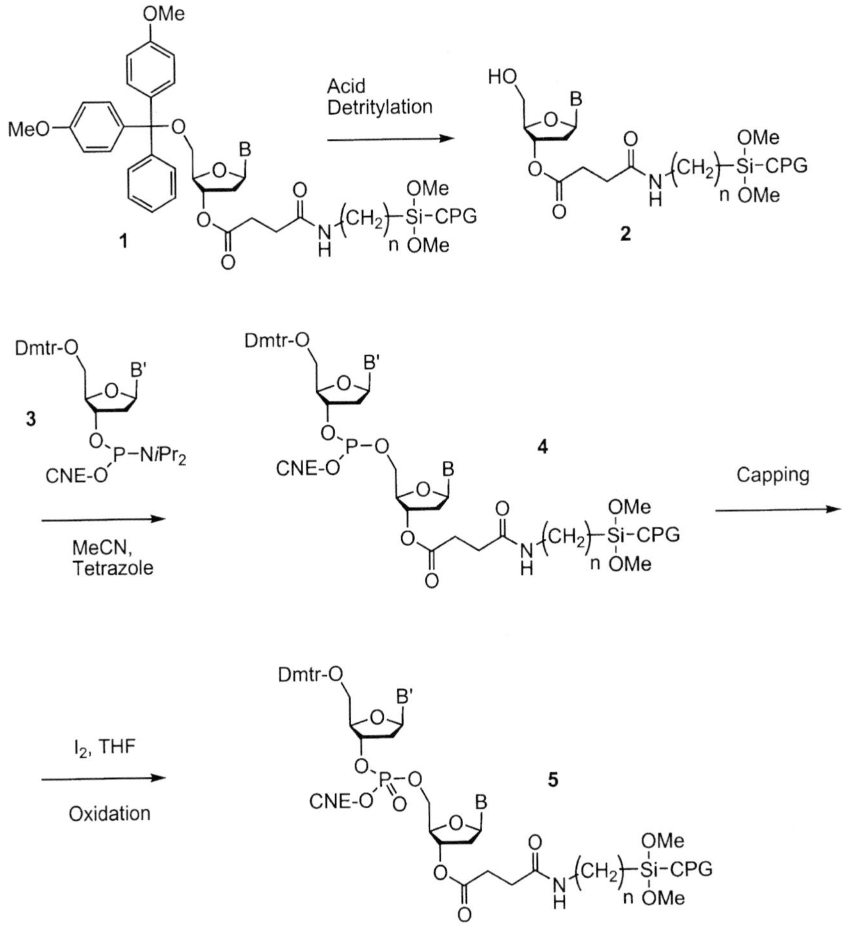

Scheme 2.5 Scheme of the oligonucleotide synthesis on a DNA synthesizer.

ing block **3** which is protected at the 5'-hydroxy group (and on the nucleobases) is coupled. On the 3'-hydroxy group, it carries a phosphorylation synthon.

The phosphoamidites are rather stable against hydrolysis and oxidation and therefore can be purified by silica gel chromatography and stored for a longer period. The alkoxy group R'O of the phosphoamidate **3** serves as a protecting group (Scheme 2.3). A weak acid like tetrazole activates the nucleoside-3'-phosphitamide and the dialkylamino group can be substituted by a 5'-OH-nucleoside (e.g. **2**).

The remaining free hydroxyl groups are capped by acetylation. For the alkyl amino leaving group diisopropyl has been shown to be a good substituent. The phosphitetriester **4** is oxidized by an aqueous iodine solution to the phosphotriester **5**. The whole reaction sequence can be carried out at room temperature, where the condensation times on the solid phase are in the range of a few minutes because of the high excess of reagents.

To summarize, there is a reaction cycle for the synthesizer, in which the oligonucleotide chain, bound to the solid support, is elongated stepwise in the 5'-direction.

1. Cleavage (detritylation) of the 5'-OH protecting group
2. Activation of the phosphoamidite by mixing it with tetrazole and addition to the solid phase-bound oligonucleotide
3. Capping
4. Oxidation of the phosphitetriester to the phosphotriester by addition of I_2.

Once the target oligonucleotide is complete, all protecting groups are cleaved and the oligonucleotide is released from the solid support in the final step.
The 5'-Dmt-group is cleaved by an acid. The following treatment with aqueous ammonia cleaves the cyanoethyl group and the acyl-protecting groups of the nucleobases. The succinate linker is also cleaved.

Purification is achieved by exclusion chromatography on Sephadex G25 to separate the oligonucleotide from small molecules such as the cleavage products of the protecting groups.

2.3.3
Hybridization with Synthetic Oligonucleotides

2.3.3.1 Principles
The hybridization of an oligonucleotide is the process by which two single strands bind together sequence selectively and form a double strand having the double helix structure (Fig. 2.2) [4].

This process is reversible and can be seen as association/dissociation or renaturation/denaturation. The stability of a duplex molecule against thermal denaturation depends on the

- *number of nucleobases*

The Watson–Crick base pairs have different stabilities. A GC base pair is formed by three hydrogen bonds, an AT base pair by only two (Scheme 2.1). Therefore the association constant of an AT pair in chloroform is approximately 100 L/mol whereas for a GC pair the value is 10^4–10^5 L/mol. That means, the more GC pairs there are in a duplex, the more stable it is.

• *sequence of nucleobases*

Duplexes with the same number of nucleobases but in a different sequence can differ in stability. Not only is the number of GC pairs important, but also their position. The dissociation of a short-length uplex normally begins at its termini. As a result, the introduction of a GC base pair at the termini hinders the dissociation more effectively than the introduction of a GC pair at an internal position. Also, the sequence of the purine/pyrimidine bases plays a major role in the stability of a duplex. The stacking pattern of the bases of the hydrophobic nucleobases plays a major role in the stabilization of a duplex. The stacking energy of the following duplexes as calculated by quantum mechanics is: 5′-C-G-3′/3′-GC-5′ duplex, 14.6 kcal/mol; 5′-G-C-3′/3′-C-G-5′ duplex, 9.7 kcal/mol; and 5′-G-G-3′/3′-C-C-5′ duplex, 8.3 kcal/mol.

• *length of the duplex*

The greater the interaction between the bases via hydrogen bonds and stacking pattern, the more stable the duplex (approximately 3–6 kcal/mol for each hydrogen bond).

• *concentration of cations*

Nucleic acids are polyanions and in order to form a duplex, two polyanions must be in close proximity to one another. Therefore, hybridization is hindered by coulombic repulsion. Cations are able to compensate the negative charge of the phosphodiester and this is the reason for the increased stability of a duplex at high salt concentrations.

• *temperature*

Duplex dissociation occurs at high temperatures. The point at which 50% of the duplexes are dissociated is called the melting temperature T_M. The value of T_M is an indicator of the thermal stability of the duplex.

DNA double helices have a lower molar extinction coefficient than that calculated from the sum of the bases. If the thermal denaturation of a DNA double strand is measured by UV spectroscopy, the absorption (e.g. at 260 nm) increases with increasing temperature. This effect is called hyperchromicity and originates from base stacking. In the double helix, the transition dipole moments are coupled which results in a lower extinction coefficient.

The thermal denaturation of a DNA duplex is also referred to as the "melting of the DNA". The graph, which shows the temperature-dependent UV absorption, is the so-called "melting curve" (Fig. 2.3). The melting curve has a sigmoid form. The point of inflection of this curve is a phase transition and yields the "melting temperature".

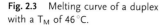

Fig. 2.3 Melting curve of a duplex
with a T_M of 46 °C.

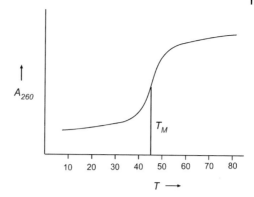

2.3.3.2 Specificity of Hybridization

The specificity of hybridization depends on the stability of the duplex and the stringency of the hybridization conditions. For small oligonucleotides (< 20 bases), the melting temperature decreases by 5–20 °C with the introduction of a single base mismatch.

At temperatures that are 5–15 °C lower than the T_M of the perfect duplex and at low salt concentrations the presence of a specific sequence can – in principle – be detected selectively. But to do this a method is needed which can detect dsDNA in the presence of ssDNA.

2.3.3.3 Dyeing Methods

dsDNA is structurally different from ssDNA. This difference can be used to selectively dye and therefore detect dsDNA [5, 6]. One special class of dyes binds selectively to duplex DNA.

The most common class of dyes used for dyeing duplex DNA binds to the duplex by intercalation i.e. the dye molecules 'push in' between the stacked bases (Fig. 2.4).

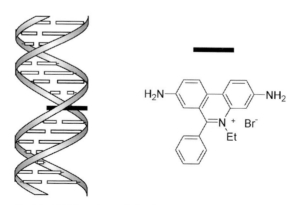

Fig. 2.4 Ethidium bromide binds to dsDNA as an intercalator.

A big advantage of using intercalated dyes is that the fluorescence of some of these dyes (e.g. ethidium bromide) is more intense than that of unbound dye. As a result, unbound dye does not have to be separated which facilitates detection of dsDNA. Even small amounts of dsDNA can be detected in the presence of ssDNA.

The principle of fluorescence intensification by intercalation is widely used especially for the identification of DNA and RNA separated by gel electrophoresis in Southern and Northern blots.

2.4
Experimental Procedures

2.4.1
Preparations

Some of the reactants are hazardous substances. In particular ethidium bromide* is very toxic, and may cause heritable genetic damage. Safety instructions for handling of the substances in use have to be followed with greatest care. Experimental work has to be carried out in an efficient hood; the use of safety glasses, a lab coat and safety gloves is mandatory!

DNA Synthesis: All the necessary solutions and chemicals are commercially available and ready to use. No further preparation is necessary.

Determination of T_M Value: A 10 mM solution of phosphate buffer containing 0.1 M NaCl and 0.1 mM EDTA* (pH=7, degassed) should be prepared and the concentrations of the DNA stock solutions should be determined.

Detection of dsDNA in the Presence of ssDNA: A 0.1 M solution of phosphate buffer containing 1 M NaCl (pH=7, degassed) and a 1×10^{-3} M solution of ethidium bromide should be prepared.

2.4.2
Synthesis and Purification of an Oligonucleotide (16 bases) and Yield Determination

One of the following oligonucleotides is to be synthesized using a DNA synthesizer

1. 5'-GGG CGC TGG AGG TGT G-3'   **6**
2. 5'-GGG CGC TGT AGG TGT G-3'   **7**
3. 5'-CAC ACC TCC AGC GCC C-3'   **8**
4. 5'-CAC ACC TAC AGC GCC C-3'   **9**

Perform the synthesis, purify the crude product and determine the yield.

---

* Hazardous chemical, see Appendix

2.4.3

Determination of the T$_M$ Value of a Match (6–8, 7–9)
and a Single Mismatch Duplex (6–9, 7–8)

2.4.3.1 **Materials**

The materials required are a stock solution of the DNA 16mers, hybridization buffer (0.1 M NaCl in 10 mM phosphate buffer and 0.1 mM EDTA pH 7 (degassed)) and cuvettes suitable for use with UV light.

2.4.3.2 **Preparation of DNA Solutions**

From the DNA stock solutions **6** or **7** (concentrations have already been determined), 1 nmol is transferred into Eppendorf tubes **a** and **b**.

To **a**, 1 nmol of oligonucleotide **8** and to **b**, 1 nmol of oligonucleotide **9** is added. Then, buffer is added to obtain a total volume of 1000 μl. The resulting concentration of the duplex is now 1 μM.

2.4.3.3 **Measurement of the Melting Temperature and Concentration of G + C**

Taking care to avoid bubbles, the DNA solutions are transferred to the cuvettes which are then placed into the thermo block of a Varian Cary 100 spectrometer. The temperature-dependent absorption of the solutions is then measured. Measurements are duplicated for heating and cooling. For details of how to use the Varian spectrometer and the software for thermal denaturation and renaturation reference should be made to the spectrometer manual. The melting temperature thus observed should then be compared to the calculated value (*http://ozone2.chem.wayne.edu/Hyther/hytherm1main.html*).

2.4.4

Detection of dsDNA in the Presence of ssDNA using Ethidium Bromide

2.4.4.1 **Materials**

The materials required are a solution of ethidium bromide 1×10^{-3} M, a 0.1 M solution of phosphate buffer pH = 7 containing 1 M NaCl, dsDNA, ssDNA, a fluorescence spectrometer and cuvettes suitable for UV and fluorescence measurements.

When using low salt concentrations, ethidium bromide can bind to DNA in two different ways. The first is by intercalation and the second by ionic interaction of the ethidium cation with the negatively-charged phosphate backbone. It is necessary to work at high salt concentrations to avoid mistakes caused by ionic interactions. To avoid non-specific interaction of ethidium bromide with the DNA, its concentration must be low. The experimenter should be aware of the physiological effects of ethidium bromide.

2.4.4.2 **Calibration**

From a DNA stock solution (not the solution of synthesized DNA), 5×1 ml calibration solutions with ODs between 0.1 and 0.8 should be prepared. The absorption of these solutions should then be measured at a wavelength of $\lambda = 260$ nm and a temperature of 80 °C. From the ODs, the exact concentration can be determined using the nearest neighbor method. After cooling to room temperature, 5 μl of ethidium bromide solution is added and the fluorescence intensity is measured at an excitation wavelength of 520 nm and an emission wavelength of 590 nm. Plotting the fluorescence intensity versus the concentration yields the calibration curve.

2.4.4.3 **Determination of the Concentration of dsDNA and ssDNA**

A test solution containing an unknown quantity of ssDNA and dsDNA is warmed to 80 °C and the OD is then measured. After cooling, 5 μl of ethidium bromide solution is added and the fluorescence at 590 nm is measured. From the OD value, the total concentration of DNA can be determined and from the fluorescence intensity the proportion of dsDNA can be calculated. The concentration of ssDNA can now be calculated by simply taking the difference.

2.4.5
Laboratory Report

A protocol of the DNA synthesis (sequence, scale, differences to the routine protocol of the DNA synthesis, trityl monitoring) should be prepared.

The yield of the synthesis should then be calculated assuming the loading on solid support to be 100%. What is the concentration of the DNA stock solution?

Determine the melting temperature of the match- and mismatch-duplex. Add the denaturating and the renaturating curves. Compare the resulting values.

Show the development of the calibration curve for the determination of dsDNA using ethidium bromide. Measure the concentration of dsDNA and ssDNA of an unknown sample.

2.5
Bibliography

1 BLACKBURN, G.M., GAIT, M.J. *Nucleic Acids in Chemistry and Biology*, Oxford: IRL Press; New York, Tokyo, 1990, 1996

2 NARANG, S.A. *Tetrahedron*, **1986**, *39*, 3–22. (Overview of DNA synthesis).

3 BANNWARTH, W. *Chimia*, **1987**, *41*, 312. (Overview of DNA synthesis).

4 SANTALUCIA, J., ALLAWI, H.T., SENEVIRATNE, P A. *Biochemistry* **1996**, *35*, 3555–3562.

5 YGUERABIDE, J., CEBALLOS, A. *Anal. Biochem.* **1995**, *228*, 208–220.

6 LEPECQ, J.B., PAOLETTI, C.J. *Mol. Biol.* **1967**, *27*, 87–106.

3

Doubly-labeled Peptide Nucleic Acids (PNAs) as Probes for the Detection of DNA Point Mutations

Olaf Köhler, Svenja Röttger, and Oliver Seitz

3.1
Abstract

The mutual recognition of two complementary nucleic acid strands is the molecular basis of any life form. It is the origin for most of the current approaches in oligonucleotide-based diagnostics and therapy. Utilization of synthetic compounds in biological assays can be advantageous because of their higher stability against chemical and enzymatic degradation and in the case of Peptide Nucleic Acids (PNA) their higher affinity and selectivity in binding DNA and RNA compared to natural nucleotides. The most commonly used hybridization assays employ solid phase in order to facilitate the separation of bound from unbound analytes. However, these heterogeneous assays are characterized by tedious washing protocols and non-specific adsorption. In contrast, homogeneous assays, e.g. molecular beacons, circumvent these problems. Because of their sensitivity and selectivity they have become a useful tool in hybridization assays.

In this practical course the fluorescence labeling of one of the most important DNA mimics, the peptide nucleic acids (PNAs), will be learned. Single-stranded PNA undergoes a conformational change while binding to complementary DNA. This structural reorganization can be spectroscopically detected if the PNA is labeled with a communicating system of fluorescent dye and quencher. As a result the former "dark" PNA strand fluoresces upon hybridization. The students will learn to know the selectivity of the PNA-based molecular beacons towards point mutation in single-stranded DNA oligomers.

3.2
Learning Targets

- Base mutations
- DNA detection
- Hybridization
- Molecular beacons
- PNA probes

Chemical Biology: A Practical Course.
Edited by Herbert Waldmann and Petra Janning

- Real-time measurements
- Fluorescence labeling
- Fluorescence spectroscopy
- Fluorescence resonance energy transfer

3.3
Theoretical Background

3.3.1
Basics of PNA

The mutual recognition of two complementary nucleic acid strands forms the molecular basis for an array of applications in both diagnostics and therapy. In the recent past much work has been devoted to establishing methods for the detection of particular nucleic acid sequences in hybridization assays.

Before continuing with this chapter the principles described in Chapter 2 should be read first.

In this practical course molecular beacons based on Peptide Nucleic Acids (PNA, Scheme 3.1) [1, 2], which are analogous to DNA, are synthesized and analyzed in biophysical experiments. In PNA the complete ribose-phosphate backbone is replaced by aminoethylglycine [3]. Despite this alteration the binding between PNAs and their complementary DNA oligomer takes place with high specificity and a higher affinity. The higher stability of a PNA–DNA complex is illustrated for example, by an increase in melting temperature of up to 1.5 K per base pair in comparison to a DNA–DNA complex.

It is important to be aware of the fact that PNAs are, despite their name, neither acids nor peptides. Being totally artificial molecules they are stable against enzymes such as proteases and nucleases.

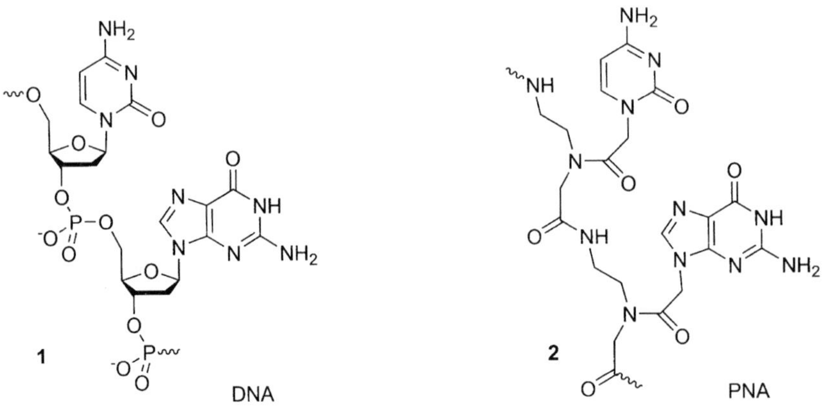

Scheme 3.1   Structure of DNA and PNA.

The structure of PNA–DNA duplexes is similar to that of DNA–DNA or DNA–RNA double strands. Nevertheless important differences exist in the single strand molecules. Single strand DNA shows only a low increase in extinction at 260 nm with rising temperature. This indicates that even in a single strand nucleobases communicate with each other and through reciprocal action the transition dipole moment lowers their total extinction coefficient. With rising temperature nucleobases show greater agility and the coupling of their transition dipole moment is suspended. Analyzing the temperature-dependent UV absorbance of single-stranded PNA oligomers, the curve shows a sigmoidal gradient comparable to the behavior of dsDNA(see Chapter 2). This indicates a phase transition between a higher ordered state at lower temperatures and a "melted" form at higher temperatures. This hypochromicity provides evidence for the more favorable base-stacking in single stranded PNA. Binding of single strand PNA to DNA must therefore lead to a conformational change which results in the well-known helical conformation of PNA–DNA duplexes [4]. Choosing two appropriate positions (e.g. the strand ends) in the PNA oligomer the distances should vary in single strand and double strand. Using fluorescence resonance energy transfer (FRET) one can visualize this conformational change. For this assay a fluorescence donor and acceptor with overlapping emission and absorption spectra are required. After exciting the donor by radiation the energy can either be transferred radiation-free to the acceptor (if they are in close proximity) or the system may fluoresce. The ratio of the energy transfer is proportional to $1/r^6$ (r=label distance) which clearly shows that the success of FRET depends on the distance between the donor and acceptor. Collisions between the fluorescence donor and the fluorescence acceptor provide an alternative means at depleting the excited state of the donor and, hence, of quenching donor fluorescence. During the hybridization process the distance between the donor and accep-

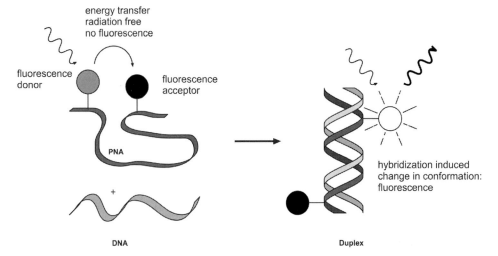

Fig. 3.1 Principle of molecular beacons.

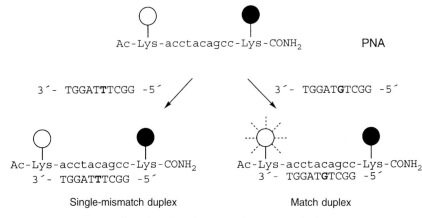

Ac-Lys-acctacagcc-Lys-CONH₂   PNA

3´- TGGATTTCGG -5´   3´- TGGATGTCGG -5´

Ac-Lys-acctacagcc-Lys-CONH₂   Ac-Lys-acctacagcc-Lys-CONH₂
3´- TGGATTTCGG -5´   3´- TGGATGTCGG -5´

Single-mismatch duplex   Match duplex

Scheme 3.2   Sequences of match and single mismatch DNA–PNA duplexes.

tor is increased so that energy transfer and collisional quenching is less likely to occur and donor emission can be detected (Fig. 3.1).

Structural probes that fluoresce after binding are called molecular beacons. DNA beacons have been known since 1996 (for more information see, *www.molecular-beacons.org*).

These double-labeled PNA oligomers will be used for the detection of a single nucleotide polymerphism (SNP). The SNPs are of special interest because it is estimated that 3000–4000 diseases are caused simply by these point mutation (so-called genetic diseases). In the case of more complex diseases like e.g. cancer or Alzheimer's disease, the likelihood of succumbing to these conditions is determined by the nature of the genetic material. In order to detect a point mutation the molecular beacon must bind to the exact complementary strand.

In this practical course the group will analyze whether a PNA beacon fluoresces only while binding to the complementary strand (match case) or also when it binds to a strand with a single mutation (mismatch) (Scheme 3.2).

3.3.2
Synthesis of PNA

First, a double-marked PNA oligomer is synthesized via Fmoc solid phase strategy (Scheme 3.3). The Fmoc group, in this case the temporary protecting group [5] for primary amino functions, can be quantitatively removed under basic conditions with DMF/piperidine. Other functional groups (for example other amino functions of the bases A, C and G) must be protected with a permanent protecting group, which is stable under all reaction conditions during the solid phase synthesis without hindering it. At the end of the synthesis it must be removable under mild conditions. A possible group, established in PNA synthesis, is the benzhydrylcarbonyl-(Bhoc-) protecting group which can be removed with trifluoroacetic acid.

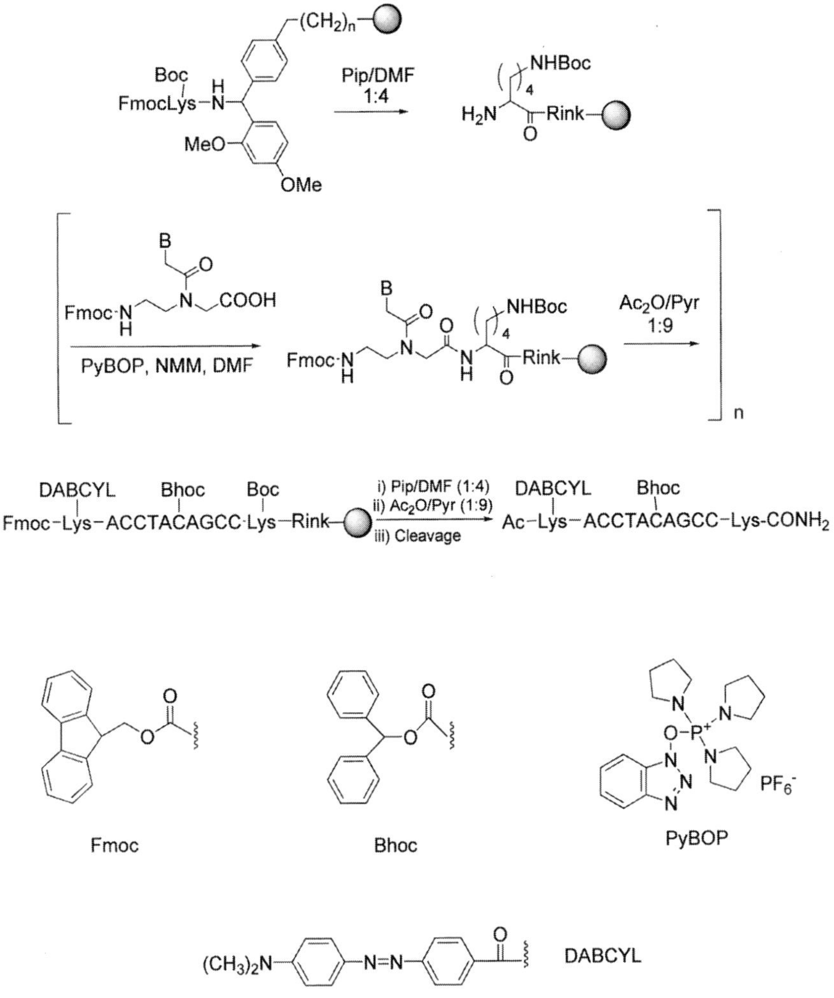

Scheme 3.3 Synthesis of the PNA probe.

After deprotection of the aminoethyl group under basic conditions (cleavage of the Fmoc protecting group) the subsequent monomer coupling is carried out with so-called coupling reagents. In the corresponding peptide synthesis a large variety of these reagents is widely used (e.g. EDC/HOBt, DIC/HOBt, HBTU, HATU) but in this case PyBOP (Scheme 3.3) is the reagent of choice as it is known to be one of the most potent reagents for solid phase peptide coupling. The advantage of Py-BOP over other reagents is that it prevents the formation of so-called guanidinium salt adducts observed with uronium-type coupling reagents such as HBTU or HATU. In addition, it is less toxic than BOP. Of course, not all of the coupling reactions will proceed with 100% efficiency.

Therefore, one cycle of PNA synthesis is not complete until unreacted amino functions have been capped in a reaction with acetic anhydride. After completion of the coupling reactions the PNA has to be cleaved off the resin. The rink-linker (Scheme 3.3) is cleaved under acid conditions (e.g. with trifluoracetic acid (TFA)) while forming the amide at the C-terminus. During this acidic treatment the permanent protecting groups (Bhoc) are also removed, so that a totally deprotected PNA strand is released into solution.

For the synthesis of doubly-marked PNA a DABCYL (as a fluorescence acceptor) carrying lysine is bound to the N-terminus in the last step before cleavage from the resin. The fluorescence donor (Scheme 3.4) is introduced afterwards in solution. For this reason a lysine has been coupled to the resin at the C-terminus, because its amino function can be selectively derivatized even in the presence of hydroxy-, carboxy- or mercapto-groups. After completing the synthesis of the PNA oligomer its hybridization and selectivity is tested with its complementary DNA and a single nucleotide mismatch strand. In the perfect case, duplex formation with a complementary strand will increase the fluorescence, whereas duplex formation will not take place with the mismatch DNA strand.

3.4
Experimental Procedures

3.4.1
Preparations

The Bhoc-protected PNA sequence FmocN-ACCTACAGCC-Lys(Boc)-resin coupled to a solid support (Rink-Amide on Tentagel) was presynthesized. The synthesis was carried out according to the following procedure.

Loading of the Resin: A solution containing Fmoc-Lys(Boc)-OH (0.1 M, 4 eq. based on the loading of the resin), PyBOP (4 eq.) and N-methylmorpholine* (NMM, 6 eq.) in DMF is added to a Rink-Amide Tentagel resin (e.g. 0.21 mmol/ g). After shaking overnight the solution is removed by filtration and the resin is washed five times with DMF. One ml Ac_2O/Pyr (1:9) is added to the resin which is then shaken for 5 min. The solution is removed by filtration and the procedure repeated once. The resin is washed with DMF (three times) and dichloromethane (five times). Loading of the resin is determined via a standard procedure [6].

Fmoc Deprotection: Pip/DMF* (1:4) is added to the resin which is shaken for 3 min. The solution is removed by filtration and the procedure repeated twice. After deprotection the resin is washed with DMF (three times), dichloromethane (three times) and DMF (three times).

Coupling: The Fmoc-b(Bhoc)-OH (4 eq., final concentration 0.1 M) building block is dissolved in DMF. N-methylmorpholine* (6 eq.) and PyBOP (4 eq.) are added. The resulting solution is vortexed for 1 min before addition to the resin.

---

* Hazardous chemical, see Appendix

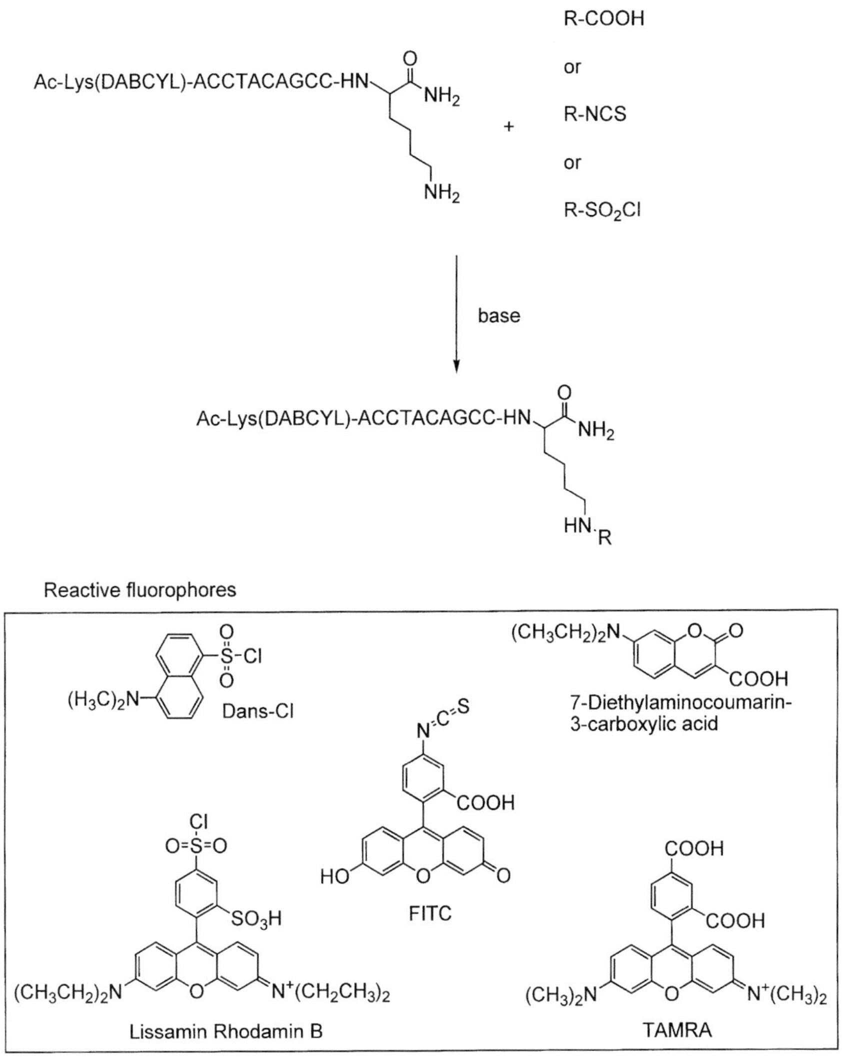

R-COOH

or

R-NCS

or

R-SO₂Cl

Ac-Lys(DABCYL)-ACCTACAGCC-HN $\ldots$

Scheme 3.4 Introduction of the florescent donor to the PNA probe.

After 1.5 h of shaking the resin is washed with DMF (three times), dichloro-methane (three times) and DMF (three times).

Capping: Ac₂O/Pyr* (1:9; 2×5 min) is added to the resin. Afterwards the resin is washed with DMF, dichloromethane and DMF (three times each).

Final release from solid support: Before cleavage the resin is washed with dichloromethane (five times). The solid support is then treated three times with approximately 0.5 mL cleavage solution (50 mg H-Cys-OMe in 10 mL TFA/m-cresol*/water (93/5/2)) for 30 min each time. The combined filtrates are concentrated under reduced pressure. The product precipitates upon addition of cold diethyl ether*. The precipitate is collected by centrifugation and removal of the supernatant. The solid is dissolved in an appropriate solvent (e.g. 10% acetonitrile/water*) and purified first by solid phase extraction using a SepPak cartridge. The red-colored eluent obtained after addition of 25% acetonitrile/0.1% TFA/74.9% water* is collected and concentrated *in vacuo*. The final purification is carried out by preparative RP-HPLC. Yields are determined according to Beer's law [7].

3.4.2
Synthesis of the PNA Beacon

3.4.2.1 Solid Phase Synthesis of the PNA–DABCYL Conjugate
The groups get the pre-synthesized Bhoc-protected PNA sequence FmocN-ACCTACAGCC-Lys(Boc)-resin coupled to the solid support (Rink-Amide on Tentagel). In the last step of the solid phase synthesis the groups have to couple Fmoc-protected and DABCYL-labeled lysine before cleavage of the resin. Final purification will be carried out by preparative HPLC.

(a) Deprotection (Cleavage of the Fmoc protecting group)

After swelling in DMF the pre-synthesized PNA oligomer **7** is treated twice with 20% piperidine in DMF (about 0.5 ml) for 5 min. After deprotection the resin is washed three times with each of the following: DMF, DCM and DMF.

(b) Coupling of Fmoc-Lys(DABCYL)OH

Four equivalents (in relation to the loading of the resin) of the amino acid derivate, 6 eq. NMM and 4 eq. PyBOP are dissolved in DMF (final concentration of amino acid derivate 0.1 M) and immediately transferred to a syringe. The suspension is then shaken for 90 min followed by the washing procedure described in (a).

(c) Deprotection, Capping

The resin is again treated twice with 20% piperidine/DMF* for 5 min. After the washing procedure (as described in (a)) two aliquots of 10% acetic acid anhydride in pyridine (about 0.5 ml) is transferred to the reactor (over 5 min for each addition) to cap the N-terminus of the oligomer. The resin is finally washed three times with DMF and five times with DCM.

(d) Cleavage

To release the labeled PNA oligomer the solid support is treated with a solution containing 50 mg HCys-OMe in 10 ml TFA/m-cresol/water* (93:5:2 v/v/v) three times for 30 min. Without neutralization 4/5 of the solution is distilled under reduced pressure. The oligomer is then precipitated by addition of cold diethyl ether. After centrifugation and discarding of the ether layer the solid is dissolved in an appropriate amount of water (about 100 µl/µmol).

The DABCYL-labeled oligomer is then detected by MALDI-TOF mass spectrometry and purified by HPLC.

3.4.2.2 Fluorescence Labeling of Ac-Lys(DABCYL)-ACCTACAGCC-HN-Lys-NH$_2$

To determine the yield of the solid phase synthesis 10 µl of the prepared solution is added to 990 µl methanol* in a cuvette. After measurement of the zero value the absorbance at 428 nm is recorded from which the concentration of the solution is calculated ($\varepsilon = 30000$ M^{-1} cm^{-1}).

Solution A: In a 500-µl reaction tube 150 µl of the prepared solution is evaporated to dryness. To solvate the residue a solution containing 0.1 M DIPEA* in 40% acetonitrile/water (v/v) is prepared and added so that the final concentration of 7 is 4 mM (if solid remains additional pure DIPEA is added; the amount of DIPEA required is dependent on the total volume).

Solution B: In another tube the reactive fluorophore is diluted to give a 0.16 M concentration in dry DMF. The volume needed for reaction is 1/4 of solution A.

Solution B is added in four portions (each 1/16 volume of solution A) over a 30-min period (e.g. if solution A is 40 µl then 2.5 µl of solution B is added four times). During the reaction time (1 h) the vessel is occasionally vortexed.

To clean up the mixture a SepPak-C18 cartridge is activated (5 ml methanol) and equilibrated (5×1 ml water) and kept on water until the reaction mixture is transferred to it. Then the cartridge is washed with appropriate amounts of water, 10% acetonitrile/water, 20% acetonitrile/water and 40% acetonitrile/water. For better elution the latter three should contain 0.1% TFA. Via MALDI-TOF analysis the product-containing fractions are determined. To check their purity analytical HPLC (and if necessary a preparative HPLC) is carried out.

The yield is determined as described above but with water as the solvent and absorbance measured at 260 nm ($\varepsilon = 93400$ M^{-1}cm^{-1}).

3.4.3
Detection of Single Nucleotide Polymorphism via Fluorescence Spectrometry

A buffer containing 0.1 M NaCl, 10 mM Na$_2$HPO$_4$ and 0.1 mM EDTA in water (pH = 7.0) is prepared. Before utilization the buffer is degassed for about 15 min under reduced pressure.

Fluorescence measurements are carried out at room temperature at a concentration of 1 µM of DNA and PNA. First, the signal of the buffer is measured (zero value). Second, the PNA solution is added and the signal is measured again. Third, the DNA solution is added, the solution is warmed up to approx. 60°C (water bath) and cooled down to room temperature again to speed up duplex formation. The fluorescence is measured a third time and the changes in the signal are observed.

3.5
Bibliography

1 NIELSEN, P.E., EGHOLM, M., BERG, R.H., BUCHARDT, O. *Science* **1991**, *254*, 14971500. (First article about PNAs).

2 HYRUP, B., NIELSEN, P.E. *Bioorg. Med. Chem.* **1996**, *4*(1), 5–23.

3 UHLMANN, E., PEYMAN, A., BREIPOHL, G., WILL, D. *Angew. Chem. Int. Ed.* **1998**, *37*, 2796–2823.

4 KUHN, H., DEMIDOV, V.V., COULL, J.M., FIANDACA, M.J., GILDEA, B.D., FRANK-KAMENETSKII, M.D. *J. Am. Chem. Soc.* **2002**, *124*, 1097–1103.

5 GREENE, T.W., WUTS, P.G.M. *Protective Groups in Organic Synthesis*, 3rd edn **1999**, Wiley-Interscience (pp 272, 506–509, 513, 521, 532, 540).

6 GORDEEV, M. F., LUEHR, G. W., HUI, H. C., GORDON, E. M., PATEL, D. V. *Tetrahedron* **1998**, *54*, 15879–15890.

7 WALLACE, R. B., SHAFFER, J., MURPHY, R. F., BONNER, J., HIROSE, T., ITAKURA, K. *Nucleic Acids Res.* **1979**, *6*, 3543.

4

Synthesis and Characterization of a Covalent Oligonucleotide–Streptavidin Conjugate and its Application in DNA-directed Immobilization (DDI)

Florian Kukolka, Marina Lovrinovic, Christof M. Niemeyer, and Ron Wacker

4.1
Abstract

Semi-synthetic DNA–protein conjugates are synthesized by covalent coupling of thiol-modified DNA oligonucleotides and streptavidin (STV). The resulting conjugates have binding capacity for four equivalents of biotin and nucleic acids of complementary sequence. The conjugates are purified to homogeneity by ultrafiltration and chromatography, and characterized by photometry and gel electrophoresis. Subsequently, the conjugates are applied as molecular linkers in the DNA-directed immobilization of a biotinylated enzyme to microplates containing complementary capture oligonucleotides.

4.2
Learning Targets

- Biotin–streptavidin interaction
- Protein crosslinking
- DNA crosslinking
- Bioconjugate purification
- Electrophoretic characterization of DNA–protein conjugates
- Microplate analyses
- Solid-phase DNA hybridization
- DNA and protein microarray technology

4.3
Theoretical Background

4.3.1
(Strept)avidin–Biotin System and DNA Oligonucleotides as Molecular Tools

Streptavidin (STV, MW approx. 56 kDa) is a homotetrameric protein isolated from *Streptomyces avidinii* which shows a remarkable and unique interaction with its

Chemical Biology: A Practical Course.
Edited by Herbert Waldmann and Petra Janning
Copyright © 2004 WILEY-VCH Verlag GmbH & Co. KGaA, Weinheim
ISBN: 3-527-30778-8

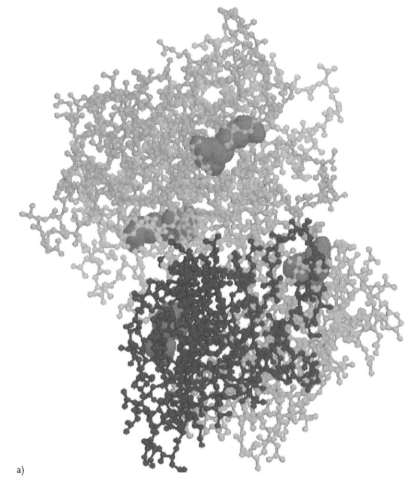

a)

b)

Fig. 4.1 Molecular structure of STV[1] (**a**) and D-biotin (**b**).

low-molecular weight ligand D-biotin. The molecular structure of STV (Fig. 4.1 a) is highly conserved and reveals a large homology to the homotetrameric protein avidin (MW approx. 60 kDa), which can be isolated in large amounts from egg white. The biomolecular recognition of the water-soluble molecule biotin (vitamin H, Fig. 4.1 b) through (strept)avidin is characterized by the extraordinary affinity constant of about 10^{14} dm^{3}mol^{-1}, indicating the strongest ligand–receptor interaction currently known [1]. Since biotinylated materials are often commercially available or can be prepared with a variety of mild biotinylation procedures, biotin–(strept)avidin conjugates form the basis of many diagnostic and analytical tests [2]. Despite its homology, STV lacks the carbohydrate moiety of avidin, and thus gives an enhanced performance in many analytical tests, e.g. significantly reduced non-specific binding. Another great advantage of STV is its extreme chemical and thermal stability. Streptavidin is resistant to many proteases under physiological conditions including proteinase K and it can be heated repeatedly at the temperatures needed for PCR cycling with no apparent damage. It survives extremes of pH and can still bind biotin, for example, even in the presence of 7 M urea.

Short DNA oligonucleotides are also powerful tools in biomedical diagnostics because of their great specificity in stringent hybridization which allows any unique DNA sequence – 16 to 20 nucleotides in a target with the complexity of a mammalian genome (ca. 3×10^9 bp) – to be detected specifically and, in principle, in isolation. The power of DNA as a molecular tool is enhanced by the procedures currently available to synthesize virtually any DNA sequence by automated methods, and to amplify any DNA sequence from microscopic to macroscopic quantities by the polymerase chain reaction (PCR). Another very attractive feature of DNA is the great rigidity of short double helices (30 to 60 base pairs), so that they behave effectively like a rigid rod spacer between two tethered binding sites on both ends.

4.3.2
Covalent DNA–Streptavidin Conjugates

The covalent attachment of an oligonucleotide moiety provides STV with a specific binding domain for complementary nucleic acids in addition to its four native binding sites for biotin. This bi-specificity of the hybrid molecules allows them to serve as universal, efficient and highly selective connectors in the oligonucleotide-directed assembly of proteins and other molecular and colloidal components [3]. The covalent attachment of the oligonucleotide to STV is carried out with the aid of a hetero-bispecific crosslinker, sulfosuccinimidyl-4-(N-maleimidomethyl)cyclohexane-1-carboxylate (sSMCC, Scheme 4.1). The addition of thiols to maleimides yields stable thioether conjugates. This crosslinking method is chosen because 5′-thiol modified oligonucleotides can be prepared on a standard DNA synthesizer. The ε-amino groups of Lys side chains of STV are first derivatized with an sSMCC crosslinker to provide a maleimide functionality, which is subsequently reacted with the thiolated oligonucleotide. The crosslinked products are pre-purified by ultrafiltration, and then fractionated by anion exchange chromatography on a Mono Q HR 5/5 column using an automated FPLC system. The latter step leads to the quantitative sep-

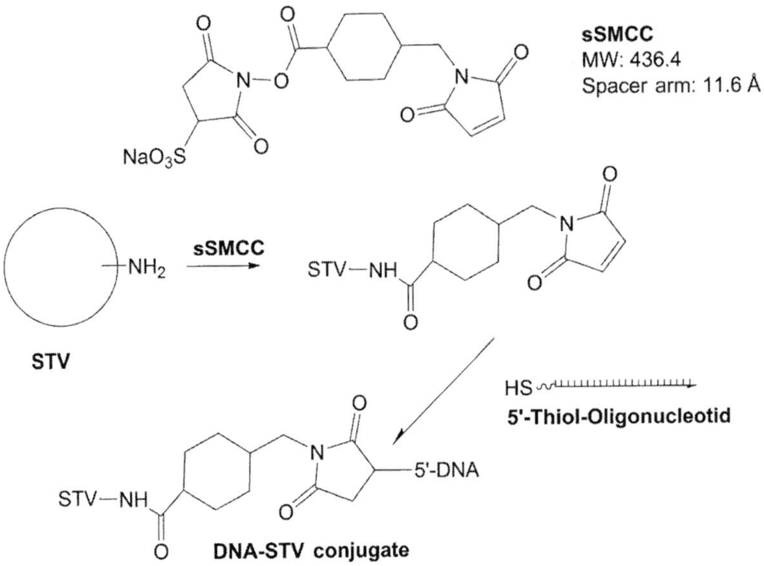

Scheme 4.1 Covalent cross-linking of STV and 5-thiolated oligonucleotides.

aration of DNA–STV conjugates containing different numbers of attached DNA strands.

4.3.3
DNA-Directed Immobilization

Subsequent to its purification and characterization, the covalent DNA–STV conjugate is employed in the DNA-directed immobilization (DDI) of proteins. DDI enables the production of laterally microstructured protein arrays. Such devices are currently of great interest due to the demands of high-throughput biomedical analysis and proteom research [4]. While comparable microarrays comprised of oligonucleotides can easily be fabricated by automated deposition techniques [5], the stepwise, successive immobilization of proteins on chemically-activated surfaces is obstructed by the general instability of sensitive biomolecules, often leading to a high tendency for denaturation. The DDI method provides a chemically mild, site-selective process for the attachment of multiple delicate proteins to a solid support (Scheme 4.2) [6]. The DDI method uses DNA microarrays as a matrix for the simultaneous, site-selective immobilization of many different DNA-tagged proteins or other molecular compounds. Since the lateral surface patterning can now be carried out at the level of the physicochemically stable nucleic acid oligomers, the DNA microarrays can be stored almost indefinitely, functionalized with proteins of interest via DDI immediately prior to use, and subsequent to hybridization, the protein arrays can even be regenerated by alkaline denaturation of the double helical DNA linkers.

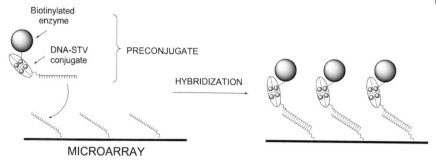

Scheme 4.2 Schematic drawing of the DDI method.

Here, a model DNA array on a microtiter plate is used as an immobilization matrix. For preparation, a microplate is coated with STV, and biotinylated capture oligonucleotides are immobilized as capture probes. Several capture probes are immobilized which are all complementary to the covalent DNA–STV conjugate but differ in length. In addition, one non-complementary capture probe is bound to the plate as a negative control, allowing for the estimation of the specificity of the DNA-directed immobilization.

Biotinylated alkaline phosphatase (bAP) is used as the protein to be immobilized. Various conjugates are prepared using different stoichiometric amounts of the bAP and the covalent DNA–STV conjugate (bAP:DNA–STV=0.2, 1.0, and 5.0). From these bAP–STV–DNA conjugates a serial 1:10 dilution is prepared leading to four final conjugate concentrations, ranging from 1 nM to 1 pM. The bAP only preparation (without coupling to the DNA–STV conjugate) is also applied as a negative control. Additional controls are prepared from bAP, a biotinylated oligonucleotide (bA24) and native STV in various stoichiometric compositions (bAP:bA24:STV=10:10:10, 1:10:10, 10:10:1). The working concentration of these control conjugates is adjusted to 1 nM. All conjugates are allowed to hybridize to the microplate-bound capture oligomers for 45 min at room temperature. Subsequent to washing, the fluorogenic substrate Attophos is added to the wells and quantification of fluorescence is carried out with a microplate reader after incubation times of 5, 10 and 20 min.

4.4
Experimental Procedures

Note: Some of the reactants are hazardous substances. In particular, acrylamide* and sodium azide* are very toxic. Acrylamide may cause cancer and heritable genetic damage. Safety instructions for the handling of these substances have to be followed with great care. Experiments should be carried out in an efficient hood only; the use of safety glasses, a lab coat and gloves is mandatory.

4.4.1
Conjugate Synthesis

4.4.1.1 Oligonucleotide Activation
To 100 µL (100 µM) of oligonucleotide A24 (sequence: TCC TGT GTG AAA TTG TTA TCC GCT) in TE buffer, pH = 7.4, add 60 µL of a 1 M DTT solution, briefly mix and incubate for 2 h at 37 °C.

4.4.1.2 STV Activation
Dissolve one aliquot (approx. 2 mg) of sSMCC in 60 µL of DMF* (shake at 35 °C if required). Add the sSMCC solution to 200 µL of a 100 µM solution of STV in PBS buffer pH = 7.3 and incubate in the dark at room temperature for 1 h.

4.4.1.3 Purification of Activated Oligonucleotide
Purify the activated oligonucleotide by gel filtration chromatography using a Superdex Peptide column (Pharmacia) connected to a Fast Protein Liquid Chromatography (FPLC) system. The column should be equilibrated for at least 10 min using the same conditions as for the purification. Inject the 160 µl of activated oligonucleotide and elute the sample with PBSE buffer, pH = 6.8, using a flow rate of 0.7 mL/min. The absorbance should be measured at 260 and 280 nm and the samples represented by peaks collected in 0.55 mL fractions. Pool the collected fractions from the main peak (elution volume approx. 9–11 mL), which contain the activated oligonucleotide. Immediately use the oligonucleotide in the crosslinking reaction (Section 4.4.1.5).

Task: Describe the chromatogram obtained and identify the oligonucleotide peak according to its retention time and to the absorbance ratio at 260 and 280 nm (compare with Section 4.4.1.7).

4.4.1.4 Purification of Activated STV
Remove the top and bottom caps of two disposable gel filtration columns (NAP5 and NAP10, Pharmacia) and pour off the conserving liquid. Support the column over a suitable receptacle and equilibrate the columns by gravity flow-through of three complete fillings with PBSE, pH = 6.8. Apply 260 µL of the activated STV just on top of the filter plate of the NAP5 column and allow the liquid to completely enter the gel bed. Adjust the sample volume to 500 µL by adding 240 µL of PBSE, pH = 6.8. Elute the activated STV with 1 mL of PBSE, pH = 6.8, and collect 1 mL of filtrate. Apply the collected filtrate sample to the NAP10 column. Elute the activated STV with 1.5 mL of PBSE, pH = 6.8, and use as soon as possible in the crosslinking reaction (Section 4.4.1.5).

---

* Hazardous chemical, see Appendix

4.4.1.5 Cross-linking of STV and Oligonucleotide, Quenching and Buffer Exchange

Mix the purified activated STV (Section 4.4.1.4) with the oligonucleotide fractions (Section 4.4.1.3) and incubate the solution for 1.5 h in the dark at room temperature.

Transfer the mixture into a molecular cut-off ultrafiltration unit (Centricon 30, Millipore) and reduce the volume to 600 µL by alternately centrifuging at 4000 r.p.m. and shaking the mixture. Add 1 µL of 1 M mercaptoethanol* and further diminish the volume to approx. 200 µL. Add 1 mL of Tris A and repeat the concentration procedure until the sample volume is about 200 µL.

4.4.1.6 Purification of STV–Oligonucleotide Conjugate

Flush and equilibrate an anion-exchange column (MonoQ HR5/5, Pharmacia, or equivalent) connected to an FPLC system with Tris buffer containing three different concentrations of NaCl (c=0, 0.3 and 1 M) by appropriately mixing the flow of two buffer solutions containing 0 M (Tris A) and 1 M NaCl (Tris B), starting with 1 M and ending with 0.3 M. A constant flow of 0.4 mL/min should be maintained. Execute each step until the detection baselines at 260 and 280 nm are constant. The final equilibration time with 0.3 M NaCl needs to be at least 10 min before purification of the conjugate can be initiated.

Transfer the concentrated sample (Section 4.4.1.5) into the sample loop of the FPLC system using an appropriate injection needle. Care should be taken to avoid trapping air bubbles in the sample loop! Maintain the salt concentration at 0.3 M and the flow at 0.4 mL/min and inject the sample onto the column. Elute the fractions with a NaCl gradient ranging from 0.3 to 1 M, as indicated in Tab. 4.1. Collect 0.55-mL fractions. Record the absorbance at 260 and 280 nm. Pool the peak fractions, exchange buffer and concentrate the conjugate samples by twofold ultrafiltration (see Section 4.4.1.5) using 500 µL of TBSE each time. Store the fractions at 4 °C until further use.

Task: Label the eluted peaks in the chromatogram with serial numbers. Discuss the elution profile and identify peaks of educts and products. Take into account the results obtained in Sections 4.4.1.7 and 4.4.2.

Tab. 4.1 NaCl gradient of anion exchange chromatography

Volume (mL)	NaCl conc. (mol/L)
0–4	0.3
4–8	0.3–0.36
8–41	0.45–0.8
41–61	1

4.4.1.7
Photometry and Conjugate Quantification

Determine the ratio of the absorbance at 260 and 280 nm of DNA (tA24), STV and three different stoichiometric mixtures of the DNA and STV (1:1, 1:2 and 2:1). Use a DNA concentration of about 1 µM for these measurements. Quantitate the concentrated peak fractions by measuring the absorbance at 260 and 280 nm. The absorbance measured arises from the absorbance of both the STV and the oligonucleotide moiety present in the sample:

$$A280 = A280_{STV} + A280_{DNA} \tag{1}$$

$$A260 = A260_{STV} + A260_{DNA} \tag{2}$$

Assume that the absorbance ratio (A260/A280) for both the isolated STV and oligonucleotide is a constant value:

$$A260_{DNA}/A280_{DNA} = \alpha \tag{3}$$

$$A260_{STV}/A280_{STV} = \beta \tag{4}$$

Insertion of Eq. (3) and (4) into (1) and (2), respectively, leads to Eq. (5) and (6):

$$A280 = A280_{STV} + (A260_{DNA} \times 1/\alpha) \tag{5}$$

$$A260 = A260_{DNA} + (A280_{STV} \times \beta) \tag{6}$$

Insertion of Eq. (6) into (5) leads to (7 a)

$$A280 = A280_{STV} + (A260 - A280_{STV} \times \beta) \times 1/\alpha \tag{7a}$$

Rearrangment of Eq. (7 a) gives the absorbance at 280 nm which is due to the STV moiety:

$$A280_{STV} = (A280 - (A260 \times 1/\alpha))/(1 - \beta \times 1/\alpha) \tag{7b}$$

Once the absorbance of STV at 280 nm is known, the concentration can be calculated using Lambert–Beer's Law:

$$c = A/(\varepsilon \times l) \tag{8}$$

The extinction coefficient (ε) of tetrameric STV at 280 nm is $\varepsilon = 142\,400\ M^{-1}\ cm^{-1}$. The concentration of ssDNA can be calculated from the approximation that at 260 nm 1 OD is equal to approximately 33 µg/mL of ssDNA.

Task: Calculate and use the absorbance ratios determined experimentally for the various mixtures of STV and TA24 to identify the peaks of the two chromatograms (Sections 4.4.1.3 and 4.4.1.6). Calculate the isolated yield of the DNA–STV conjugate HA24.

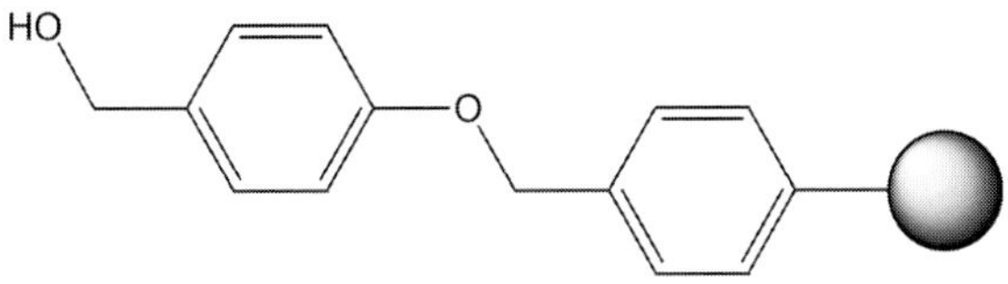

6

Scheme 5.6 Structure of the Wang linker (**6**) for solid phase peptide synthesis.

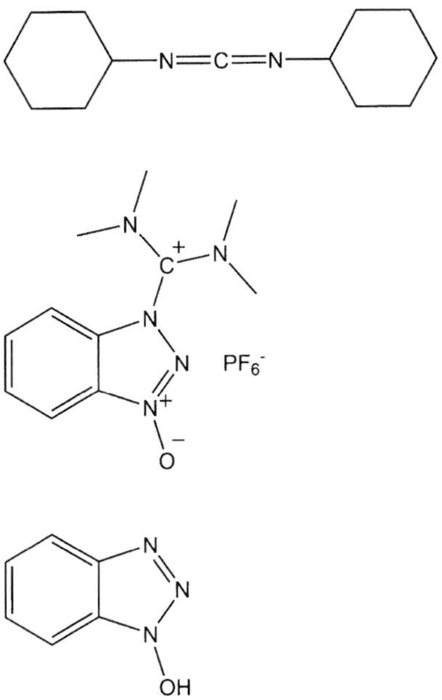

7

8

9

Scheme 5.7 Structures of coupling reagents: DCC (**7**), HBTU (**8**) and HOBt (**9**).

5.4
Experimental Procedures

5.4.1
Preparations

5.4.1.1 Cultivation of PC-12 Cells

Note

Because PC-12 cells grow at a very slow rate, it is necessary to plan this experiment well in advance.

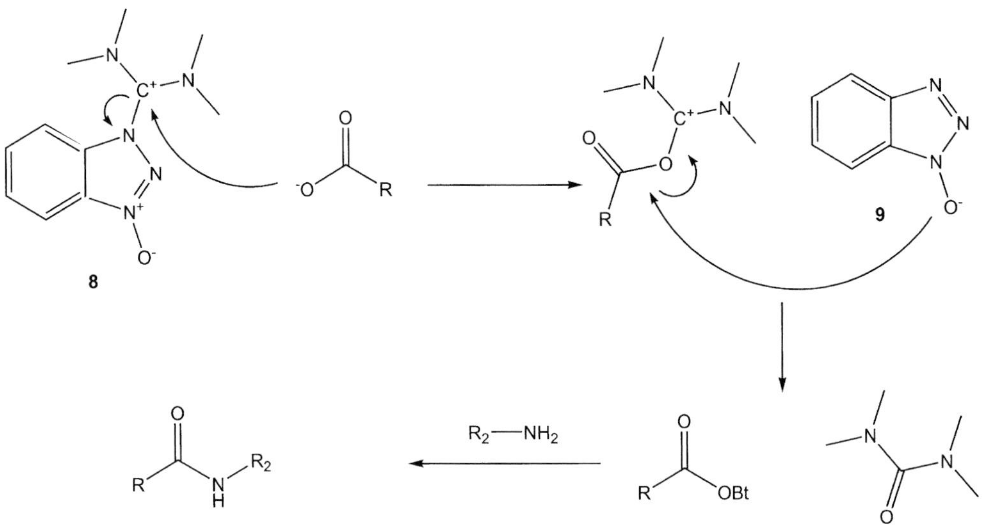

Scheme 5.8 Potential coupling mechanism using HBTU (**8**) and HOBt (**9**).

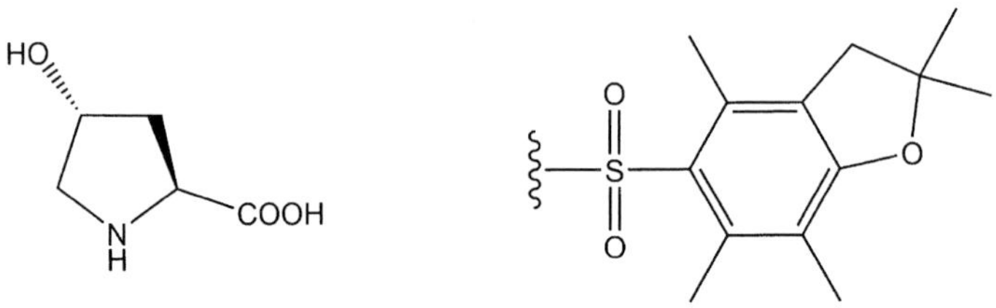

Scheme 5.9 Structures of hydroxyproline (Hyp, (**11**)) and 2,2,4,6,7-penta-methyl-dihydrobenzofurane-5-sulfonyl (Pbf) protecting group (**12**).

Materials

- PBS buffer solution: 140 mM NaCl, 2.7 mM KCl, 1.5 mM KH$_2$PO$_4$, 8.1 mM Na$_2$HPO$_4$ (pH 7.3).
- Trypsin EDTA solution: 0.05% trypsin, 0.53 mM Na$_4$EDTA*, Hank's Balanced Salt Solution (HBSS).
- Culture medium: Dulbeccos' modified Eagle with sodium pyruvate, 1000 mg/L glucose, pyridoxin, 10% horse serum, 5% fetal calf serum, 2 mM L-glutamine, 100 units/mL penicillin, 100 µg/mL streptomycin sulphate.

* Hazardous chemical, see Appendix

Cell Culture

PC-12 cells were cultivated in Falcon tissue culture flasks (area 25 cm^2) at 37 °C and 10% CO_2. The following procedure was used to transfer the cells to new flasks (transfer at 50% confluence in a ratio of 1:5):
- removal of the culture medium
- washing of the cells with 2 mL PBS buffer
- incubation with 0.5 mL trypsin EDTA solution for 3 min at 37 °C
- transfer to a centrifuge flask containing 3 mL culture medium (carefully)
- centrifugation at 600 g for 3 min
- suspension of the pellet in fresh culture medium
- distribution to five new tissue culture flasks, containing 4 mL each of culture medium.

5.4.1.2 Synthesis of the Hexameric Peptide

Materials

All Fmoc-protected amino acids were purchased from Novabiochem and used without prior purification. Dichloromethane was distilled under argon from CaH_2.

Synthesis of the Bradykinin Hexameric Peptide Gly-Phe-Ser-Pro-Phe-Arg-Wang

Fmoc-Arg(Pbf)-OH (7.18 g, 11.07 mmol) is dissolved in dry dichloromethane to give a clear, colorless solution, cooled to 0 °C and diisopropylcarbodiimide* (3.42 mL, 22.14 mmol) is added at this temperature. The solution is further stirred at 0 °C for 20 min and the solvent is evaporated under reduced pressure. The resulting white foam is dissolved in 60 mL dry DMF* and added to pre-swollen Wang hydroxy resin (3 g, 1.23 mmol/g loading) followed by DMAP* (45 mg, 0.369 mmol, added as solid). The slurry is shaken for 2 h at room temperature then washed 5 times with 100 mL DMF, and twice with dichloromethane*.

The peptide chain is elongated following standard procedures for Fmoc peptide synthesis using 4 eq. Fmoc-protected amino acid, 3.6 eq. HBTU, 4 eq. HOBt, and 8 eq. diisopropylethylamine*.

5.4.2
Peptide Synthesis

5.4.2.1 First Steps

The solid phase synthesis of bradykinin or its analogs is carried out on a 0.05 mmol scale in a solid-phase syringe reactor. Fmoc strategy and a Wang linker are used. Therefore dry polystyrene resin is provided containing a hexapeptide attached via a Wang linker (Gly-Phe-Ser-Pro-Phe-Arg-Wang and Gly-Phe-Ser-D-Phe-Phe-Arg-Wang resin). For the synthesis of these hexapeptides see Section 5.4.1. The side-chain hydroxy group of serine is protected with an acid labile *tert*-butyl group and the guanidine group of arginine with the acid-labile Pbf group. From both resins the loading is determined. Then the appropriate amount of resin

(0.05 mmol scale) is weighed out into a solid-phase syringe reactor and dichloromethane is added for 30 min to swell the resin. Afterwards the resin is washed five times with DMF. The peptide is assembled onto the resin using the coupling procedures and the peptide is then cleaved from the resin and purified.

5.4.2.2 Determination of Loading Efficiency

The dry Fmoc-protected amino acid (4–5 mg) is weighed out and 3 mL of a 20% piperidine/DMF solution is added. After 15 min, 0.5 mL of this solution is pipetted out into a 3-mL silica UV cuvette (d = 10 mm), and diluted to a total volume of 3 mL. Using a spectrophotometer the absorbance of the reference (blank piperidine/DMF solution) is first measured; the absorbance of the sample solution is then read at 301 nm. By applying Lambert–Beers' law ($A = \varepsilon \, l \, c$; A, absorption; ε, molar extinction coefficient; l, path length; c, concentration of the solution), the loading of the resin can be determined ($\varepsilon = 7800 \ M^{-1} cm^{-1}$). *Note:* if the resin has been extensively dried under high vacuum, incomplete resin swelling may lead to inaccurate values of the loading.

5.4.2.3 Coupling Procedure

The N-terminal Fmoc protecting group is cleaved with 5 mL of a 20% solution of piperidine in DMF for 10 min (the cleavage is repeated twice). The resin is then washed seven times with DMF. The amino acid (4 eq.) is dissolved in 1 mL DMF, and then pre-activated by the addition of HBTU (3.6 eq.), HOBt (4 eq.) and DIPEA* (8 eq.). After 2 min of pre-activation, the amino acid mixture is added to a slurry of the resin, and shaken for 2 h at room temperature. The resin is then washed seven times with DMF. This procedure is repeated until the complete sequence has been assembled.

5.4.2.4 Cleavage Procedure

The N-terminus is deprotected with 20% piperidine in DMF. The resin is washed seven times with dichloromethane. After transferring the resin to a round-bottomed flask, 5 mL of the cleavage cocktail is added (95% TFA*, 2.5% tri*iso*propyl silane (TIS*), 2.5% water). The slurry is shaken for 4 h at room temperature, and the resin is filtered. The filtrate is evaporated and the crude peptide is precipitated out of the resulting oil by the addition of ice-cold diethyl ether*.

5.4.2.5 Purification

The crude peptide mixture is purified by HPLC using a semi-preparative reverse-phase column (Macherey-Nagel, Nucleodur C18 gravity, 5 μ, 21 mm ID, 125 mm column length) with acetonitrile/water mixtures. Approximately 20 to 30 mg of the crude peptide is dissolved in 450 μL water, filtered and subjected to HPLC with the following elution gradient (A = water containing 0.1% TFA; B = acetonitrile* containing 0.1% TFA) at a flow rate of 15 mL/min (detection wavelength 215 nm): 0–4 min 10% B, 4–40 min linear gradient up to 40% B, 40–50 min linear gradient up to 100% B for washing and re-equilibration of the column.

The peptides elute at between 20 and 30 min after injection: for example, brady-kinin has a retention time of 21 min, and the [3Hyp,D-Phe] peptide 26 min.

5.4.2.6 Characterization of the Peptides

The peptides are characterized by analytical HPLC/ESI-MS (separation conditions similar to the preparative chromatography runs but with 0.1% formic acid as elu-ent additive instead of TFA, detection of positive ions) or by MALDI-TOF, using DHB as a matrix.

5.4.3
Measurement of Intracellular Ca^{2+} Levels

5.4.3.1 Incubation and Measurement Buffer

For calcium measurements, the cells must lay in a serum-free buffer which con-tains ions in a concentration similar to physiological conditions and some glu-cose. Buffer (100 mL): 140 mM NaCl, 5 mM KCl, 1 mM $MgCl_2$, 1 mM $CaCl_2$, 20 mM HEPES, 1 mM Na_2HPO_4, 5.5 mM glucose. Adjust the pH to 7.3–7.4.

5.4.3.2 Fura-2 Loading

Without detaching the cells from the bottom of the tray and using a transfer pi-pette, remove the medium and replace with 2 mL freshly prepared incubation buf-fer. The supernatant is once again exchanged with 2 mL incubation buffer, and 3 µL of a stock solution of Fura-2 (1 mM in DMSO) is added along with 2 µL 5% pluronic in DMSO. After incubating for at least 30 min at room temperature in the dark (*note*: once in contact with Fura-2, the cells must be protected from light) on a slowly rotating shaker, the Fura-2-containing supernatant is removed with a disposable pipette. The cells are suspended in four or five steps using a total of 2–3 mL measurement buffer. The suspension is then transferred to 15-mL Falcon tubes, and centrifuged at a maximum of 450 g (800 r.p.m., 17.5 cm rotor) at room temperature. The supernatant is discarded and the cells are carefully resuspended in 2–3 mL measurement buffer.

5.4.3.3 Calcium Measurements

Calcium is detected using a spectrofluorometer (e.g. model EB 50, Perkin Elmer). Under conditions described here, it is possible to measure changes of the calcium concentration in the cytoplasm and to understand the principle of calcium measure-ments. It is not possible to determine the absolute calcium concentration, because the baseline is not stable enough. The cell suspension is transferred to a 3 mL bo-rate glass cuvette and put in the spectrometer. The fluorescence of Ca^{2+}-bound and unbound Fura-2 is measured by rapidly alternating the dual excitation wavelength between 340 and 380 nm and electronically separating the resultant fluorescence signals at the emission wavelength of 515 nm. The ratios (R) of the fluorescence at the two wavelengths are computed and used to calculate changes in $[Ca^{2+}]_i$. First, it is necessary to wait at least 5 min in order to obtain a stable reference baseline. Then, 2 μl bradykinin aqueous stock solution or a solution of the agonist is added. The end concentration in the cell suspension should be 5 μM, the concentration of the stock solution is calculated, corresponddingly. When the baseline has stabilized again, 2 μl of an antagonist (same concentration as before) are added.

To calculate $[Ca]_i$ it is necessary to know the maximum (R_{max}) and the minimum (R_{min}) fluorescence ratios of fura-2. R_{max} is determined by adding 10 μl of the detergent NP40 to lyse the cells and R_{min} by adding 500 μl of an EDTA* solution (200 mM stock solution, pH 7.4) to complex the Ca^{2+}-ions.

5.5
Bibliography

1 GRYNKIEWICZ, G., POENIE, M., TSIEN, R. Y. *J. Biol. Chem.* **1985**, 3440–3450.
2 HAUGLAND, R. P. *Handbook of Fluorescent Probes and Research Chemicals*, 6th edn, **1996**.
3 TSIEN, R. Y. *Biochemistry* **1980**, *19*, 2396–2404.
4 (a) ATHERTON, E., SHEPPARD, R. C. *Solid Phase Peptide Synthesis – A Practical Approach*, IRL Press, Oxford, **1989**. (b) GRANT, G A. (ED.). *Synthetic Peptides – A Users Guide*, W. H. Freeman and Company, **1992**, pp. 78–148. (c) Novabiochem catalog,

2002, pp. 1.1–1.3; 2.1–2.4 and 3.1–3.19. (d) SONGSTER, M. F., BARANY, G. *Methods Enzymol.* **1997**, *289*, 126–174.
5 MERRIFIELD, R. B. *J. Amer. Chem. Soc.* **1964**, 304–305.
6 (a) CHAN, W. C., WHITE, P. D. (EDS). *Fmoc Solid Phase Peptide Synthesis: A Practical Approach*, Oxford University Press, **2000**, pp. 1–74. (b) FIELDS, B., NOBEL, R. L. *Int. J. Peptide Protein Res.* **1990**, *35*, 161–214.
7 ALBERICIO, F., CARPINO, L. A. *Methods Enzymol.* **1997**, *289*, 104–126.

6
In silico Protein Ligand Design

Marcus A. Koch, Lars Arve, Lars Kissau, and Jantje M. Gerdes

6.1
Abstract

In silico design of protein ligands and enzyme inhibitors respectively, plays a crucial role in modern drug research. Due to increased computing power and enhanced software performance, molecular modeling has become a widely used tool in modern pharmaceutical research and development processes.

In this chapter, the analytical and theoretical background is discussed to gain a better understanding of the processes involved in producing a molecular model.

The program PEP is used to design a potential peptidic ligand for a given protein, the three-dimensional structure of which is known. During this process, the CHARMm force field is used to minimize structures. After analyzing the results, the designed ligand and a non-peptidic molecule are dissected retrosynthetically.

6.2
Learning Targets

- Design of ligands/inhibitors
- Molecular modeling
- Force fields
- Minimization techniques
- *De novo* design
- Enzyme kinetics
- Pharmacokinetics
- (Q)SAR
- Retrosynthesis
- Stereoselective synthesis strategies

Chemical Biology: A Practical Course.
Edited by Herbert Waldmann and Petra Janning
Copyright © 2004 WILEY-VCH Verlag GmbH & Co. KGaA, Weinheim
ISBN: 3-527-30778-8

6.3
Theoretical Background

6.3.1
Overview

Computer-based tools have become very important in drug discovery. Methods of computational chemistry are currently routinely used for analysis of ligand–receptor complexes at the atomic level and for the calculation of properties of small molecules as potential drug candidates.

The aim of this experiment is to become familiar with modern strategies of finding and optimizing new physiologically active compounds.

The *in silico* design of potential ligands (inhibitors or activators) for biological macromolecules (usually proteins) in order to modulate their function can be outlined as follows:

- The native three-dimensional structure of the target protein should be known at atomic resolution which is accessible by means of structure analysis techniques. When only sequence information is available, a structural model can be conceived using homology modeling or threading techniques provided that the structure of a homologous protein is known.
- Analysis of the binding site will reveal the key features required to enable small molecule ligands to bind to that site. Important considerations in the design of ligands are the type of amino acids constituting the cavity and thus interacting with bound molecules, the volume of the binding site and the electrostatic potential of its surface. Sometimes known or even co-crystallized ligands can be used as starting points for further design and/or optimization.
- Virtual compound libraries are then docked into the binding pocket. Binding affinity is evaluated and an individual fitness value is assigned to each compound thus leading to a ranking list. This virtual docking procedure can either be performed in a one- or two-step process. In the two-step process, the compounds of the library are designed and then docked into the binding pocket. In the one-step procedure, the compounds are generated *de novo* by successively adding fragments to a start fragment already docked into the binding pocket. Thus, constitutive and conformational aspects of a probable binding mode are already taken into consideration at the design level. The results of such methods must be examined carefully with regard to their practicality by checking the proposed binding mode and the interactions found by the respective program. After evaluation of such first design hits *in vitro*, which usually means synthesis of the promising compounds and subsequent enzyme, toxicity and cell assays, a lead structure may be found which – after further optimization – may yield a drug candidate.

Commonly, *de novo* design begins with the identification of a biological target. If the molecular basis of a disease is known (most often a dysfunctional protein), the protein structure is determined and analyzed as accurately as possible. Even in this early stage the use of computers is absolutely imperative. The structural data has to be complemented e.g. by addition of hydrogen atoms following certain algorithms.

Experience has shown that although crystallized proteins display somewhat reduced activity compared to the native state, the activity is still sufficient to draw conclusions concerning the active site. Potential binding sites are examined for the spatial distribution of hydrogen bond donors and/or acceptors. Optimal positioning of potential counter groups is extracted from previous data (e.g. from the *Cambridge Structural Database*). Lipophilic patches are also identified as partners for hydrophobic interactions.

The next step is the actual ligand design. Commonly, two different strategies are followed: on the one hand, fragments are fitted into the designated space (*Building*) which is created in the actual molecules by the operator. On the other hand, a molecule can be formed successively by connecting fragments to an already existing structure (*Linking*). Alternatively, one can test single compounds of a database (e.g. the *Available Chemicals Database ACD*) for their binding affinity to the protein. This *in silico* equivalent of screening is called *Docking*.

One of the most preeminent problems with all of the methods described is the assessment of the structures generated by the programs. How can the best ligand be chosen from all the structures found? Since the crystal structures are rigid the flexibility of a protein can only be estimated. The appraisal or prediction of a K_i-value (inhibition constant) is very problematic. All the results from a computational method have to be taken only as suggestions. In this respect, it is mostly the operator's intuition and experience which helps to evaluate these suggestions. At this point, the synthetic feasibility should also be taken into consideration.

In cases where the structure of a target is unavailable, comparison of substances with biological activity can lead to the generation of a ligand with higher affinity. Analysis of structural similarities (if some are found) allows for conclusions about the environment in the bound state. On the basis of these conclusions, modifications of known structures can be suggested leading once again to higher affinities.

6.3.4.2 Program for Engineering Peptides (PEP)

The computational approach described herein is based on the iterative, successive design of peptidic ligands, which are built up from proteinogenic or non-proteinogenic amino acids. This progressive build-up starts with a seed fragment placed in an appropriate region of the binding site. The seed position(s) can be determined by X-ray or NMR studies of ligand-protein complexes. If no structure is available, seeds must be obtained by manual or computer-aided docking.

This build-up approach uses fragments from a user-defined library. In each growth step, a genetic algorithm is applied (see below) for conformational optimization of the last fragment added to reduce the computation times. Additionally, a special solvation model built into PEP is used to classify the binding energies of the ligands in solution to obtain more realistic results. After each growth step, all the ligands are ranked according to their binding energy and only the Top Ten ligands are used in the following growth step, thus preventing combinatorial explosion and long computation times.

The main disadvantage of this procedure is that each growth step passes off without knowledge of the following step even though the success of each growth step depends strongly on the previous steps. Therefore, the quality of the scoring functions built into PEP is extremely important for good results.

Growing Procedure
Commencing from a starting amino acid ("seed"), peptides are constructed by iteratively adding amino acids in conformations that interact most favorably with the residues in the receptor binding site. After each growth step, a certain number of sequences is kept for further growth (default value is 10). The monomeric building blocks are taken from a user-defined topology library, which contains the atom types, atomic partial charges, covalent bond lengths and a list of rotatable dihedrals for each monomer. At first, the *in vacuo* total energy (inter- and intramolecular) of the last added monomer is optimized by the genetic algorithm described below, while most of the already prepared ligand is kept rigid. The conformations of the last added monomer are then ranked according to their *in vacuo* energy and filters are applied to discard residues with internal hydrogen bonds and rotamers that would lead to further growth in a forbidden direction (e.g. the wall of the binding pocket).

After all the amino acids of the topology library have been minimized, the binding energy in solution is calculated for the best rotamer of each residue and only the highest scoring sequences are retained for the next level of growth. The program then tests whether the latter are dead ends or whether further growth or elongation would only lead to poor interactions with the receptor. This procedure is then repeated on the second growth level. Each amino acid in the library is attached to each of the 10 dipeptide sequences retained from the first step, minimized and finally scored. Successive growth levels, therefore, generate peptides that are lengthened by one amino acid until a defined peptide length is reached. The output data provided by PEP include residue sequences, energies and atomic coordinates of the peptide in the PDB-file format.

Genetic Algorithm (GA) for Ligand Conformational Search
A genetic algorithm (GA) is a stochastic optimization method that mimics the process of natural evolution by manipulating a population of data structures called chromosomes. Starting from an initial randomly-generated population of chromosomes, the GA repeatedly applies two mutually exclusive genetic operators, one-point crossover and mutation, which yields new chromosomes (children) that replace appropriate members of the population.

The data structure of the chromosomes is the following: every chromosome consists of so-called genes that encode

- the values of the angles of rotation around the N rotatable bonds of the added monomer,
- the Φ- or Ψ-dihedral angle of the preceding amino acid,
- and three angles that define the rigid body orientation of the added residues.

A chromosome of $N + 4$ genes, therefore, encodes the orientation and conformation of a residue with N rotatable bonds.

During the evolution process both genetic operators are applied to parent chromosomes randomly selected from existing populations with a bias towards the fittest. A one-point crossover creates two new chromosomes by swapping two segments of two parent chromosomes after a randomly selected gene. Mutation leads to a new chromosome by randomly flipping the bits (like base pairs in DNA) of selected genes of the parent chromosome. The crossing over and mutation operators are mutually exclusive, meaning that either one or the other of these, but not both, can be applied during the generation of a new chromosome. At each reproduction event, the operator is chosen using a roulette wheel method so that mutation and crossing over are selected with a chance of 80 and 20%, respectively.

The emphasis on the survival of the fittest introduces an evolutionary pressure into the algorithm and ensures that over time the population should move toward the minimum conformation(s). The chromosomes are first ordered by decreasing energies. If the total energy of a filial chromosome is lower than of a parental chromosome, it replaces the parental chromosome in the population.

The reason why such a sophisticated algorithm is applied is to reduce computation time since not all possible conformations of a peptide have to be generated, minimized and finally ranked.

6.3.5
Enzyme Catalysis and Inhibition

Enzymes are classified according to the type of reaction they catalyze and are assigned a so-called four digit E.C. number. The following six enzyme classes can be classified from E.C.1.x.y.z to E.C.6.x.y.z: oxidoreductases, transferases, hydrolases, lyases, isomerases and ligases (e.g. farnesyl-diphosphate farnesyltransferase (FTase) E.C.2.5.1.21).

In order to influence enzyme activity, knowledge of the catalytic mechanisms of enzymes is required. Known mechanisms of catalysis are acid–base, covalent, metal ion and electrostatic catalysis as well as catalysis through proximity and orientation effects or through stabilization of the transition state conformation.

Enzymes are biological catalysts that accelerate reactions by lowering the activation energy of the rate-determining step. This occurs via the formation of an intermediate complex between enzyme and substrate. The acceleration of reaction velocity also depends on the enzyme concentration in solution.

Enzyme activity is regulated in various ways. Aside from controlling the activity via regulation of its biosynthesis and biological degradation, the conformation of an enzyme can be altered by effectors which, once bound, modulate enzyme activity. Such effectors can decrease enzymatic activity. They are then called inhibitors. Naturally-occurring inhibitors are involved in the control of metabolic and signal transduction pathways. Artificial inhibitors might be used as tools for studying biological processes or may be used as drugs. Many toxic compounds are enzyme inhibitors. They are toxic because they disrupt vital processes in which enzymes or receptors are involved. Inhibitors can interact with an enzyme in different ways and enzyme kinetics is a major tool used to distinguish between these mechanisms.

Enzyme inhibitors can be roughly divided into *reversible* and *irreversible* inhibitors. Reversible inhibitors form weak bonds with the enzyme which are formed rapidly but also break easily. In consequence reversible inhibitors do not permanently disable the enzyme. Irreversible inhibitors, also known as enzyme inactivators or suicide substrates, interact with the enzyme by forming strong, usually covalent bonds. Since the enzyme–inhibitor complex formed is extremely stable, the enzyme is permanently disabled. As covalent bonds are formed slowly compared to the formation of weak bonds, irreversible inhibition displays time dependency, which means that the degree of inhibition increases with the time during which the enzyme is in contact with the inhibitor. Known examples of such suicide substrates are inhibitors of acetylcholine esterase such as *Sarin*, a nerve gas, or *E 605*, a pesticide.

The following enzyme inhibitors are of the reversible type. Each inhibitor type shows a characteristic kinetic profile. For a detailed representation of the underlying kinetics (*Michaelis–Menten*) the appropriate textbooks (e.g. Voet and Voet) should be consulted.

- A *competitive* inhibitor is a compound which competes with the substrate for binding to the active site. When bound to the active site in place of the substrate, it blocks the active site so that the substrate cannot be converted to the product. This is possible because the inhibitor molecule shares structural similarity with the substrate. The inhibition can be overcome by increasing the substrate concentration.
- A *noncompetitive* inhibitor binds to an allosteric site on the enzyme – remote from the active site – inducing a conformational change in the active site. A classical *noncompetitive* inhibitor has absolutely no effect on substrate binding but decreases the conversion velocity of bound substrate to product. Adding more substrate cannot reverse the inhibitory effect. A *noncompetitive* inhibitor binds to both free and substrate-bound enzyme.
- An *uncompetitive* inhibitor is incapable of binding to free enzyme, it can only bind to the enzyme–substrate complex. This can be explained by the fact that the inhibitor binding site only becomes available after a conformational change induced by substrate binding. The enzyme-bound inhibitor lowers the conversion rate of the substrate by a conformational change at the active site.

6.3.7.3 **Drug Optimization**

Since a drug influences pathophysiological processes and is itself influenced by the nature of the organism, optimization has to take place in two dimensions:

- Improvement of the pharmacodynamic properties should ideally lead to more potent (lower dose is required for obtaining the desired effect) and more selective (fewer side-effects) but less toxic drugs. Starting from a lead compound in the drug discovery process that already displays the desired biological effect, different optimization strategies are possible. If the central structural motif, the so-called pharmacophore, is known, systematic variation of substituents permits deduction of significant structure–activity relationships (SAR). Molecular modeling plays a major role in this optimization process. Systematic optimization approaches are homologation which means successive elongation of alkyl substituents by methylene groups, chain branching, ring–chain transformations and bioisosteres. Substituents subsumed under bioisostery have similar biological effects such as O/S, Br/iPr, H/F, $CO_2H/SO_3H/PO_3H_2$/tetrazole, $C=O/C=S$.
- Improvement of the pharmacokinetic profile of a drug comprises higher metabolic stability as well as enhancement of water solubility and membrane permeation which are both determined by the lipo- and hydrophilic properties of the respective compound, which makes optimization a tightrope walk. As a rule of thumb "Lipinski's Rule of *Five*" can be applied to describe the properties of "drug-like" molecules [1]: poor absorption or permeation are more likely when there are more than *five* hydrogen bond donors and more than 10 (2×5) hydrogen bond acceptors in the compound and when the $\log P$ (partition coefficient, see below) value is > 5 and the molecular weight is > 500 Da.

6.3.7.4 **Quantitative Structure–Activity Relationships (QSAR)**

In contrast to SAR, which aims at making as many analogs as possible, testing them and thus establishing trends, QSAR describes structure–activity relationships using quantitative methods. The biological response (IC_{50}, EC_{50}, LD_{50} to name some examples) is correlated with physicochemical (=quantifiable) properties which depend on the chemical structure. Using mathematical models, prediction of properties and activities for congeneric compounds (compounds belonging to the same structural class) is possible. One important physicochemical descriptor used is lipophilicity, represented by the partition coefficient (P), usually expressed as $\log P$, which measures the propensity of a compound to partition itself between an aqueous and a non-aqueous phase. The water/n-octanol system is usually used in order to measure P. The n-octanol mimics the nature of membrane lipids. P is defined as follows (Eq. 6.10):

$$P = \frac{[\text{compound}]_{n\text{-oct}}}{[\text{comound}]_{aq}(1-a)} \tag{6.10}$$

a: degree of dissociation from ionization constant

Amongst other physicochemical parameters usually considered are the molar refractivity as a polarizability and steric parameter and electronic parameters such as *Hammett's* σ constant.

Classical *Hansch* analysis pioneered by *Hansch* and co-workers in the 1960s pulls together structure and activity in a simple mathematical model that allows quantitative predictions of activity in complex systems. It is assumed that the sum of the substituent effects on the steric, electronic and hydrophobic interaction of the respective compound with the structure of the biological target determines the biological response in a linear free energy relationship.

6.3.8
Retrosynthesis

The concept of retrosynthesis was developed by E.J. Corey in 1957. In this approach the structure of the *synthetic target* is subjected to a deconstruction process which corresponds to the reverse of a synthetic reaction, so as to convert that target structure to simpler precursor structures, without any assumptions with regard to starting materials. Each of the precursors so generated is then examined in the same way and the process is repeated until simple or commercially available structures result. The transformation of a molecule into a synthetic precursor is accomplished by the application of a *transform*, the exact reverse of a synthetic reaction, to a target structure. The resulting intermediates of a transform do not have to be realistic molecules but theoretical constructs or *retrons* or *synthons*, which allow further transforms. These retrons have to be translated into real *reagents* during the course of the synthesis. A schematic view of retrosynthesis is outlined in Fig. 6.4.

As an example of this concept, the retrosynthesis of estrone (**1**) by Vollhardt in 1977 is shown in Scheme 6.1.

When planning a synthesis, the use of protecting groups should be taken into consideration. Since not all functional groups are inert to all reaction conditions, the use of protecting groups can change the sequence of envisaged synthetic transformations.

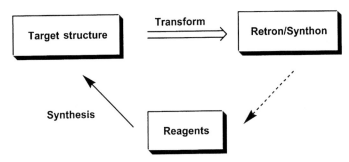

Fig. 6.4 Concept of retrosynthesis.

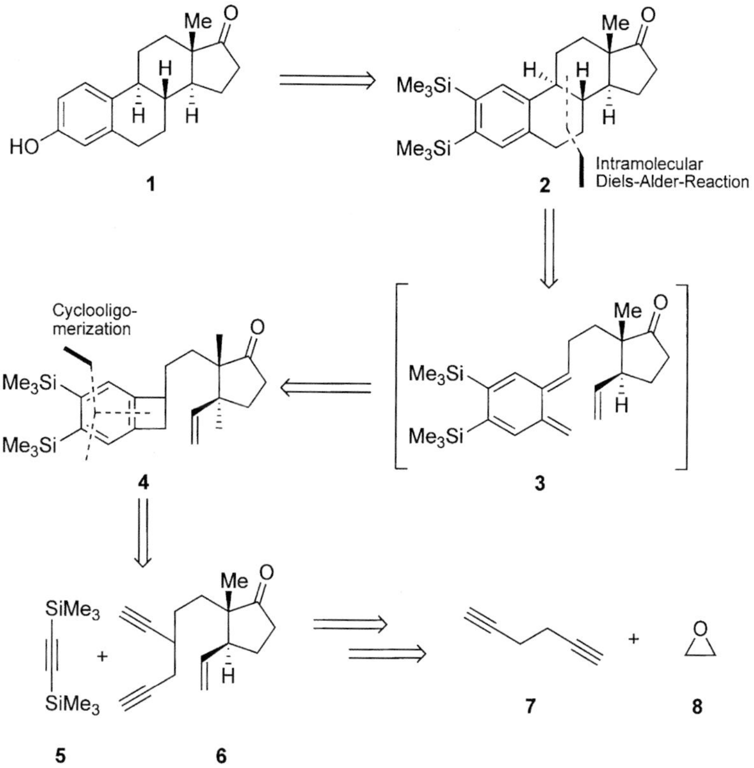

Scheme 6.1 Retrosynthetic analysis of estrone (**1**) (K. P. C. Vollhardt, 1977).

6.4
Experimental Procedures

6.4.1
General Remarks

The computational part of the experiment requires access to either an SGI workstation running IRIX 6.5 or higher, or a PC (cluster) running Linux. A PEP version for the desired operating system can be obtained free of charge for academic institutions from Professor Dr. A. Caflish, University of Zürich, Switzerland. Force field minimizations can be carried out by any of the presently used programs suitable for minimization of proteins. The authors prefer the use of CHARMm. However, free of charge shareware should suffice for the purposes of this experiment.

For the graphical display of protein structures and computation results, software packages like Insight (Accelrys Inc.), WitnotP (Novartis AG), SwissPDB-Viewer or any other viewing software which is able to interpret the PDB-file format can be employed. The authors make use of WitnotP which can be obtained free of charge for academic institutions [1] from the Novartis AG.

The following descriptions are used for the practical course at the International Max-Planck Research School in Dortmund and the Bioorganic practical course at the University of Dortmund, Germany, and may be adopted to the local configurations (cf. Appendices). Graphic manipulations refer to the menus in WitnotP, corresponding commands are found in all other widely-used programs. For a better view all commands that have to be entered into a command-line interface (CLI) are typed in a `different font`. Entries in the menus of WitnotP are written in *italics*.

6.4.2
The General Modeling Process

The following section describes each step of the modeling process (cf. the flow-chart in Fig. 6.5). The first step is to download the crystal structure of the protein from the Brookhaven Protein Data Bank (www.rcsb.org/pdb). Before this data can be visualized, all additional information found in the PDB file should be removed. In particular all the water molecules need to be erased. This new PDB file, which only contains the atomic coordinates of the protein, can be loaded into a viewing software package. Here, structural changes can be performed, e.g. cutting out co-crystallized ligands. At this point, hydrogen atoms should be added to the protein, and the new structure should be saved as a new PDB file.

With this new file at hand the protein structure can be minimized using force field calculations, e.g. CHARMm.

Simultaneously, the seed molecule should be prepared: if a peptidic ligand was co-crystallized in the original data of the protein, this peptide can be used. If this is not the case, a seed amino acid will need to be docked manually or computer-aided into the binding site using docking software. (Free of charge for academic institutions are e.g. AutoDOCK (www.scripps.edu/pub/olson-web/doc/autodock) and DOCK (www.cmpharm.ucsf.edu/kuntz)). Prior to any action with PEP, the structure of the seed molecule will have to be modified. PEP-specific nomenclature will then need to be adopted in the naming of the atoms of the seed molecule (cf. Scheme 6.2).

After minimization of the protein, the binding site must be defined to obtain a list of participating amino acids. This list of residues must be entered into the PEP.INP file. Additionally, some other entries must be inserted into the PEP.INP file (see below). Now PEP can calculate the potential energy of the binding site.

After this is done and some small modifications in the PEP.INP file are made, PEP can start to generate peptides, dock them into the binding site and rank them according to their energy.

---

[1] Contact Armin Widmer (armin.widmer@pharma.novartis.com) for a copy of WitnotP. A License Agreement will need to be signed.

Retrosynthesis

- WARREN, S. *Organic Synthesis. The Disconnection Approach*, Wiley, Chichester, **1982**.
- FUHRHOP, J., PENZLIN G. *Organic Synthesis: Concepts and Methods*, VCH, Weinheim, **1994**.
- NICOLAOU, K.C., SORENSEN, E.J. *Classics in Total Synthesis*, VCH, Weinheim, **1996**.

6.5.2
Special Literature

1 WALTERS, W.P., MURCKO, A., MURCKO, M.A. *Curr. Opin. Chem. Biol.* **1999**, *3*, 384–387.
2 BELL, I.M., GALLICCHIO, S.N., ABRAMS, M., BEESE, L.S., BESHORE, D.C., BHIMNATHWALA, H., BOGUSKY, M.J., BUSER, C.A., CULBERSON, J.C., DAVIDE, J., ELLIS-HUTCHINGS, M., FERNANDES, C., GIBBS, J.B., GRAHAM, S.L., HAMILTON, K.A., HARTMAN, G.D., HEIMBROOK, D.C., HOMNICK, C.F., HUBER, H.E., HUFF, J.R., KASSAHUN, K., KOBLAN, K.S., KOHL, N.E., LOBELL, R.B., LYNCH, J.J., JR, ROBINSON, R., RODRIGUES, A.D., TAYLOR, J.S., WALSH, E.S., WILLIAMS, T.M., ZARTMAN, C.B. *J. Med. Chem.* **2002**, *45*, 2388–2409.
3 ALLEN, F.H., KENNARD, O. *Chemical Design Automation News* **1993**, *8* (1), 1 and 31–37.

6.6
Appendices

6.6.1
Appendix A

Using WitnotP, a protein file can be prepared for CHARMm minimization with the following commands. Within the intranet of the MPI in Dortmund, these commands can be automatically executed by typing:

- `<"|prca molecule_name"` and

- `<"|pdbh2c molecule_name"`

These will write out the required `molecule_name.pdb` and `molecule_name.psf` files to the working directory. The script files must be placed in the same directory as the application (CHARMm or WitnotP, respectively).

Script name: `prca`

```
#!/bin/sh

echo "modify atom lig current $1 done "
echo "atom charge auto $1 done"
echo "atom type auto charmm $1 done"
echo "done"
echo "atom q $1 done MPEOE"
```

```
echo "atom name $1 -done -auto seq"
echo "resnum $1 1"
echo "done"
echo "label type $1"
echo "write mol2 $1 $1"
echo "write xpdb $1 $1"
echo "CHARMm psf $1 $1"
```

Script name: pdbh2c
```
#!/bin/sh

echo "copy $1 /ph2c_save_$1"
echo "modify atom lig current $1 done"
echo "atom charge automatic $1"
echo "atom name $1 -done -auto hheav"
echo "atom type auto charmm $1 done"
echo "done"
echo "atom q $1 done MPEOE"
echo "done"
echo "write mol2 $1 $1"
echo "write xpdb $1 $1"
echo "CHARMm psf $1 $1"
```

The procedure for minimization will vary depending on the local network set-up. Within the intranet of the MPI in Dortmund, minimization can be carried out by typing minall_run mol_name at the Unix command line. If your workstation is not set up to automatically find these necessary files on the network, create the files minall_run and charm_temp3.inp as specified below (path should be substituted by the correct path for your system). Please consult the manuals and ask your system administrator for installation instructions for the programs described.

Script name: minall_run
```
#!/bin/sh

case $# in

1)  sed s/XXXX/"$1"/" /path/charmm_templ3.inp > $1.inp
    charmm < $1.inp > $1.out 2>&1
    (echo "witnotp host none < EOF";
    echo "set gnu off";
    echo "read mol $1.mol";
    echo "read coord xpdb "$1"_m.pdb $1 -done";
    echo "modify mol name * "$1"_m";
    echo "done";
    echo "write mol2 "$1"_m" $1"_m";
```

This must be set to "`set vdW.accessType r`*" after the first PEP run.*

```
display all
ENERGY_END
#GROWING
#set param.length global 4
#set param.seed global 12345
#set param.growingMode global C
#display all
#run
#GROWING_END
```

After the first PEP run the #s must be removed throughout the whole GROWING *section.*
```
exit
```

The required parameter and topology files as well as a manual are included with a copy of PEP.

6.6.3
Appendix C

Sample PDB file

```
REMARK XPLOR coordinate file generated by WITNOTP
ATOM   1 N    CYS 1    3.460 -7.686 -1.311 1.00 49.01
ATOM   2 H    CYS 1    2.970 -8.467 -1.793 1.00  0.00
ATOM   3 CA   CYS 1    2.454 -6.796 -0.651 1.00 52.10
ATOM   4 HA   CYS 1    2.982 -6.006 -0.138 1.00  0.00
ATOM   5 C    CYS 1    1.618 -7.572  0.395 1.00 54.07
ATOM   6 O    CYS 1    1.692 -8.805  0.462 1.00 56.56
ATOM   7 CB   CYS 1    1.523 -6.202 -1.720 1.00 47.93
ATOM   8 HB1  CYS 1    0.737 -5.652 -1.224 1.00  0.00
ATOM   9 HB2  CYS 1    2.097 -5.530 -2.341 1.00  0.00
ATOM  10 SG   CYS 1    0.755 -7.478 -2.795 1.00 42.54
ATOM  11 HG   CYS 1    0.200 -8.050 -1.898 1.00  0.00
ATOM  12 C1   CYS 1    4.209 -6.932 -2.268 1.00  0.00
ATOM  13 O4   CYS 1    5.086 -7.474 -2.920 1.00  0.00
ATOM  14 C5   CYS 1    3.913 -5.455 -2.472 1.00  0.00
ATOM  15 H51  CYS 1    4.573 -5.056 -3.228 1.00  0.00
ATOM  16 H52  CYS 1    2.888 -5.334 -2.790 1.00  0.00
ATOM  17 H53  CYS 1    4.068 -4.926 -1.543 1.00  0.00
ATOM  18 N2   CYS 1    0.685 -6.904  1.243 1.00  0.00
ATOM  19 H3   CYS 1    0.626 -5.879  1.160 1.00  0.00
ATOM  20 C6   CYS 1   -0.104 -7.589  2.220 1.00  0.00
ATOM  21 H61  CYS 1   -0.733 -6.880  2.737 1.00  0.00
ATOM  22 H62  CYS 1   -0.722 -8.328  1.731 1.00  0.00
ATOM  23 H63  CYS 1    0.548 -8.078  2.929 1.00  0.00
END
```

6.6.4
Appendix D

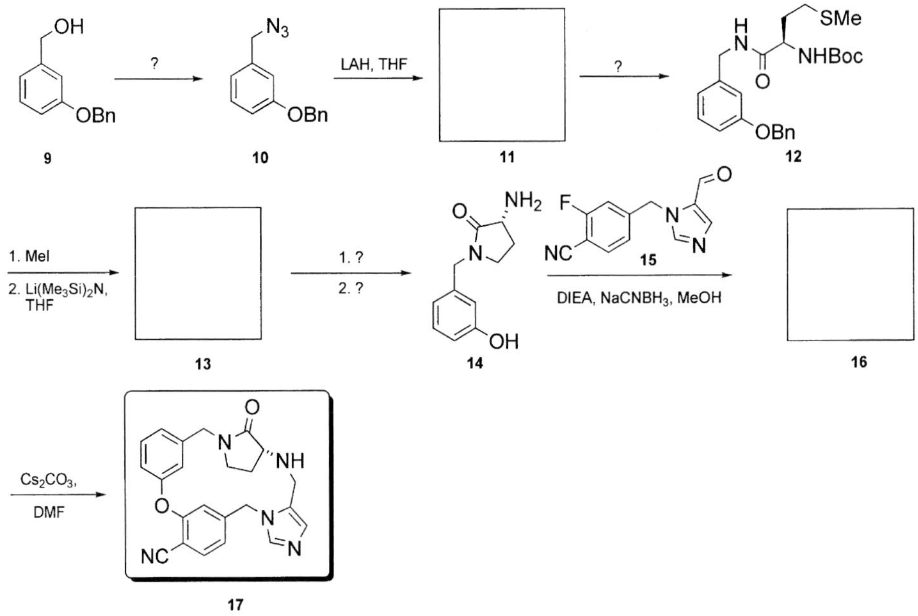

Scheme 6.3 Synthetic worksheet.

7

Lipidation of Proteins and Peptides: Farnesylation of the Ras Protein

Stefan Sommer, Tobias Voigt, Ines Heinemann, Boriana Popkirova, Reinhard Reents, and Jürgen Kuhlmann

7.1
Abstract

Lipidation is a decisive posttranslational modification for membrane-bound proteins. Ras as object of current biological and medicinal investigations is such a prenylated and membrane-bound protein. The aim of this experiment is the synthesis of differently lipidated Ras proteins and the observation of the substrate specificity of farnesyl transferase (FTase) towards different lipids when transferred to Ras. To this end, different lipid pyrophosphates will be synthesized, characterized by NMR and transferred to Ras by FTase. The Ras analogs will then be purified and characterized by SDS-PAGE and MALDI mass spectrometry.

7.2
Learning Targets

- Synthesis of farnesyl pyrophosphates and analogs
- NMR spectroscopy
- Working under anhydrous conditions
- Ion-exchange chromatography
- Biological and medicinal importance of Ras
- Mechanistic insight into lipidation of proteins
- Application of biochemical techniques: enzymatic transfer of lipid pyrophosphates
- Extraction with detergent
- SDS-PAGE
- MALDI mass spectrometry

7.3
Theoretical Background

7.3.1
Lipidation of Proteins

After translation, proteins must reach their native conformation by folding, formation of disulfide bonds and in some cases arrangement of subunits. However, many of the proteins may not have reached their active state by this stage. Further enzymatic transformations called posttranslational modifications must be carried out. Proteolytic cleavage is one of the most common posttranslational modifications, e.g. cleavage of the N-terminal hexapeptide of the inactive proprotein trypsinogen by enteropeptidase in the duodenum to release the active proteolytic enzyme trypsin. The attachment of small residues to functional groups of certain amino acids or to terminal amino and carboxylic functions is another type of posttranslational modification. Many different side-chain modifications are known including acetylation, glycosylation, hydroxylation, methylation, nucleotidylation, phosphorylation and ADP-ribosylation as well as combinations of these.

Newly expressed proteins which are eventually targeted to the membrane often undergo lipidation as one of their posttranslational modifications. This includes the attachment of three different lipids: the N-terminal myristoylation of glycines, the S-palmitoylation of cysteines and the S-prenylation of cysteines close to the C-terminus with the isoprenoids farnesyl (C-15) or geranylgeranyl (C-20) (Scheme 7.1).

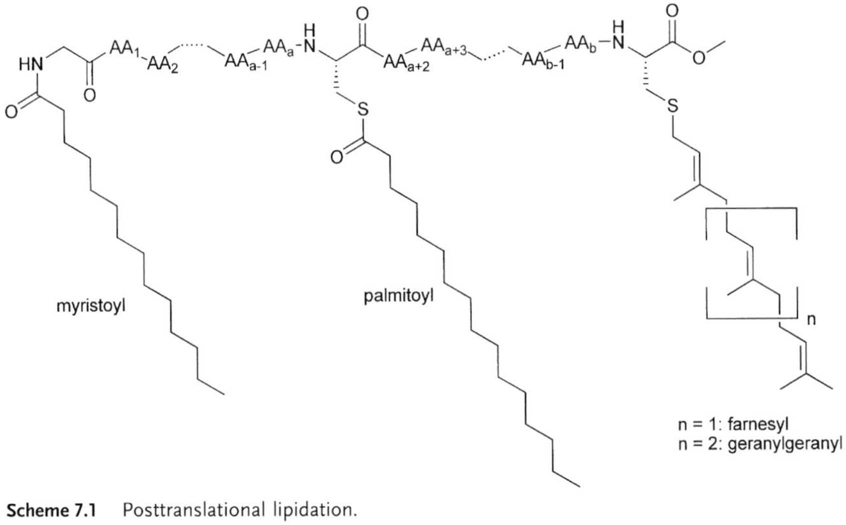

Scheme 7.1 Posttranslational lipidation.

The posttranslational modification of proteins is an important field since it has been discovered that approximately 2% of all cellular proteins are S-prenylated close to the C-terminus. These proteins play an important role in signal transduction, (e.g. Ras proteins) or intracellular transport. These hydrophobic modifications guarantee membrane localization which is essential for their biological activity.

7.3.2
Ras Proteins

7.3.2.1 Signal Transduction Pathway

Ras proteins [1] belonging to the Ras (rat-adeno-sarcoma) superfamily serve as binary switches in signal transduction pathways (Scheme 7.2).

The first step in the Ras signal transduction cascade is the dimerization of monomeric receptor tyrosine kinase achieved by the binding of an extracellular ligand, e.g. a growth factor. The monomeric subunits phosphorylate each other so that intracellular adaptor proteins will recognize a binding site and bind to the tyrosine kinase. Grb2 is the first adaptor molecule which itself will bind a second one known as Sos. In this manner the cytosolic proteins Grb2 and Sos are fixed on the inner side of the membrane and arranged in the correct order. Sos makes contact with GDP-carrying inactive Ras which is thus stimulated to exchange GDP with GTP and become active. Thereafter, active Ras can interact with effectors: proteins that interact specifically with the GTP-bound state and transmit a signal through, e.g. Raf, which is the first protein kinase in the MAP-kinase sig-

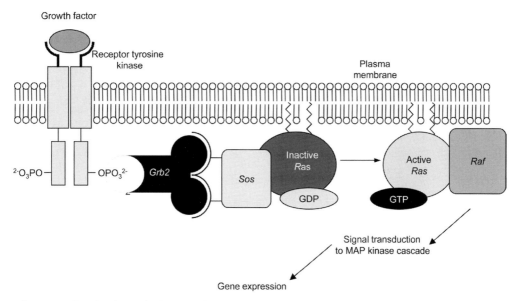

Scheme 7.2   Ras signal transduction cascade.

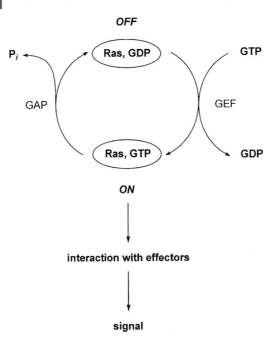

OFF

P$_i$

Ras, GDP

GTP

GAP

GEF

Ras, GTP

GDP

ON

interaction with effectors

signal

Scheme 7.3 Schematic representation of Ras regulation and function.

naling cascade leading to phosphorylation of several transcription factors and finally resulting in the expression of different genes.

Consequently, a signal from outside the cell is transferred via the Ras signal transduction cascade into the nucleus.

The signal transfer is regulated by the GTPase activity of Ras. Since the intrinsic GTPase activity is very low, the process needs to be facilitated by guanine nucleotide exchange factors (GEFs) and GTPase-activating proteins (GAPs). In its inactive state, Ras binds to GDP. Ras activation by GEF catalyzes the dissociation of GDP thus facilitating the loading of GTP. Only in this active state Ras can interact with effectors. The signaling is terminated by the GAP-aided GTPase reaction of Ras (Scheme 7.3).

An artificially extended lifetime of the signal due to an extended GTP-bound active state, resulting for example from a point mutation whose biochemical consequence is to render the protein unable to hydrolyze GTP, may lead to unregulated biological responses, such as uncontrolled cell proliferation and thereafter cancer. Activating Ras-mutations are particularly associated with myeloid malignancies and carcinomas of the colon, pancreas, lung and thyroid, but they have also been detected in other cancer types. Point mutations in the Ras oncogenes are found in ca. 30% of all human cancers.

7.3.2.2 Ras Activation

When completely modified, all four Ras isoforms, H-Ras, N-Ras, K-Ras$_A$ and K-Ras$_B$, have a farnesylated cysteine methyl ester at the C-terminus [2]. Furthermore, N-Ras and K-Ras$_A$ are palmitoylated once, whereas H-Ras is palmitoylated twice. K-Ras$_B$ contains eight lysines near its C-terminus which mediate electrostatic interactions for stable binding to the membrane. Unmodified Ras proteins are recognized by a cytosolic farnesyl transferase (FTase) at their C-terminal end CaaX (C=cysteine, a=aliphatic amino acid, X=serine, methionine, alanine, glutamate). After recognition the FTase transfers the farnesyl to the cysteine. Finally, a protease belonging to the endoplasmic reticulum (ER) cleaves the terminal tripeptide aaX whereupon the prenylcysteine carboxyl methyltransferase (pcCMT) transfers a methyl group from SAM (*S*-adenosyl-*L*-methionine) to the terminal and farnesylated cysteine (Scheme 7.4).

The next step includes the palmitoylation of Ras but the location of the palmitoyl transferase (PalTase) is still under debate. The kinetic-trapping-model [3] suggests that membrane insertion is reversible as long as the proteins are only farnesylated and becomes irreversible when they are palmitoylated. Therefore, the PalTase should be located on the cytosolic side of the membrane. A second theory [4] hypothesizes that palmitoylation takes place within the ER whereupon the completely modified protein is then transported to the plasma membrane. Such a pathway could utilize the cytoplasmic surface of secretory vesicles (Golgi apparatus) or a novel transport system.

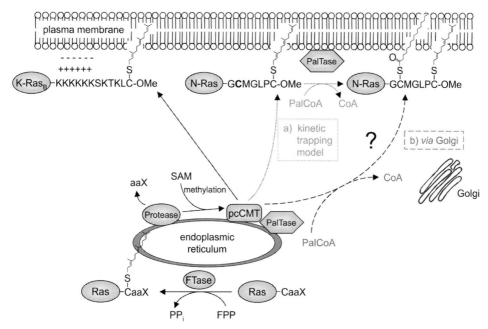

Scheme 7.4 Posttranslational modifications of Ras proteins.

The investigation of the Ras pathway and of its posttranslational modifications is an ongoing field of research. Being aware of the important role that Ras plays in different cancer types a detailed insight is necessary to find new strategies for fighting Ras-related cancer. In particular, the synthesis of lipidated Ras peptides and entire neo-Ras proteins in addition to the analysis of their membrane-binding properties, have led to a better understanding of the molecular details that govern plasma membrane targeting and the binding of lipidated proteins [5].

7.3.2.3 Farnesylation

The farnesylation of Ras proteins is essential for their biological activity. If the CaaX-cysteine is mutated to serine or farnesylation is inhibited, the Ras proteins cannot anchor themselves to the membrane. Furthermore, proteolysis, methylation or palmitoylation will not occur without farnesylation. This makes the inhibition of FTase interesting for medicinal studies and a main target for cancer treatment [6]. At the moment, at least four FTase inhibitors are in the advanced stages of clinical trials.

FTase consists of an α- and a β-subunit. Data from crystallographic studies of FTase show a barrel-like structure with a hydrophobic pocket in its center which can accommodate the farnesyl pyrophosphate (FPP). A very refined crystal structure of K-Ras$_B$ together with an FPP-based FTase inhibitor and FTase itself revealed that the CaaX motif of the peptide substrate binds in an extended conformation to the active site of FTase and directly interacts with the farnesyl isoprenoid [7]. A complementary van der Waals surface formed by K-Ras$_B$ and the isoprenoid that contacts the C-terminal aX renders FTase specific for K-Ras$_B$.

The zinc ion in the active site is required for catalysis and it is with this ion that the cysteine residue of the CaaX motif coordinates. Removal of the zinc ion results in a dramatically different peptide conformation, which suggests that a key role of the zinc ion in protein prenylation is to correctly orient the cysteine thiol(ate) for catalysis and to stabilize a productive peptide substrate conformation [7].

Finally, the FTase catalyzes farnesyl transfer to the protein via an S_N1 mechanism (Scheme 7.5).

The biosynthesis of farnesyl pyrophosphate (C-15) [3] is part of the isoprenoid metabolism. One ATP unit is consumed when the pyrophosphate residue is introduced into a phosphorylated precursor of FPP. (The biosynthesis of geranyl pyrophosphate (C-10) and geranylgeranyl pyrophosphate (C-20) is achieved in the same way). The hydrolysis of the pyrophosphate results in a strongly negative free standard enthalpy (ΔG^0) so that FPP becomes an activated form of farnesol that can be easily transferred to the protein by the FTase.

As an alternative to enzymatic farnesylation, short, farnesylated peptides can be synthesized by chemical synthesis and then linked to a C-terminally truncated protein. The connection between the synthetic farnesylated peptide and the protein can be achieved by an attached linkage group [8] or by intein-promoted protein ligation [9].

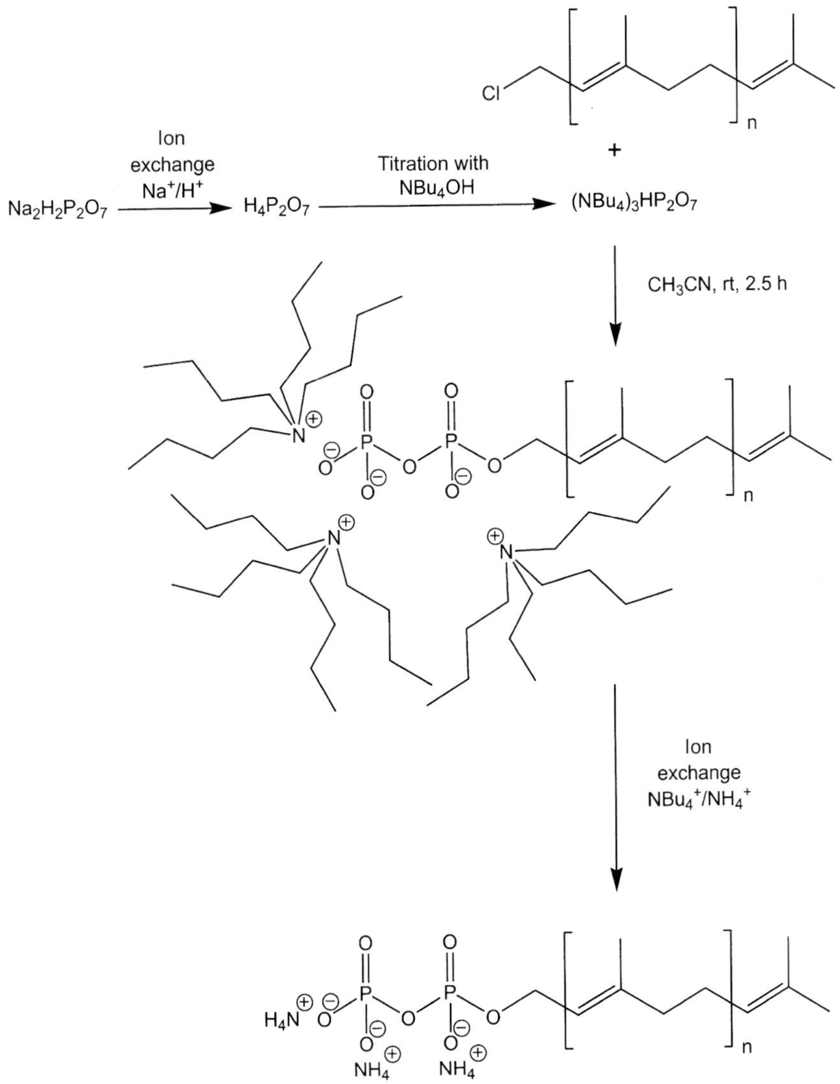

Scheme 7.8 Synthesis of isoprenoid pyrophosphates.

7.4.2.2 Preparation of the Ion Exchange Column

To regenerate the DEAE ion exchange column (20 mL), it is first washed with 250 mL of 1 M aqueous NaOH* and deionized water until the eluate is neutral. It is then washed with 250 mL 2 M aqueous HCl* and again with deionized water until the eluate is neutral. Finally, the ion exchanger is in H⁺-form again.

To obtain the NH₄⁺-form, an additional run with 25% aqueous ammonia* (approximately 13 M) and deionized water is necessary.

* Hazardous chemical, see Appendix

7.4.2.3 **Preparation of Prenyl- and Alkylpyrophosphates**

The following chemicals are used: geranylchloride (MW: 172.2 g/mol; ρ: 0.931 g/mL), farnesylchloride (MW: 240.82 g/mol; ρ: 0.916 g/mL), dodecylchloride (MW: 204.78 g/mol; ρ: 0.867 g/mL), $(n\text{-}Bu_4N)_3HP_2O_7$ (MW: 902.34 g/mol).

A solution of 3.33 g (15.0 mmol) of disodium dihydrogen pyrophosphate in 15 mL of 10% aqueous ammonium hydroxide is passed through a 2.5×7.0 cm (58 meq.) column of Dowex AG 50W-X8 cation exchange resin (100–200 mesh, hydrogen form). The free acid is eluted with 110 mL of deionized water and the resulting solution (pH 1.2) is immediately titrated to pH 7.3 with 40% (w/w) tetra-*n*-butylammonium hydroxide. The resulting solution (approx. 150 mL total volume) is dried by lyophilization to yield approximately 13.1 g of a hygroscopic white solid (97%): ^{1}H-NMR (90 MHz, D_2O) δ 0.83-1.01 (36 H, m, CH_3), 1.14-1.73 (48 H, m, CH_2), 3.03-3.19 (24 H, m, CH_2), 4.65 (s, OH); ^{31}P-NMR (32 MHz, D_2O) δ 1.60 (s).

To a well-stirred solution of tris(tetra-*n*-butylammonium)hydrogen pyrophosphate $(n\text{-}Bu_4N)_3HP_2O_7$ (2 eq.) in 1 mL of dry acetonitrile*, 80 mg of the corresponding prenyl- or alkylchloride (1 eq.) is added dropwise. The mixture is stirred at ambient temperature for 2.5 h. Afterwards the solvent is removed under vacuum. The residue is dissolved in 2 mL ion exchange buffer (*i*PrOH/25 mM $NH_4HCO_3 = 1/49$) and loaded onto the ion exchange column. At a flow rate of 1 mL/min the column is eluted employing twice its volume of ion exchange buffer. Turbidity in the eluate indicates the elution of the product. If turbidity does not occur, the fractions that contain the product can be detected by thin layer chromatography (a potassium permanganate solution is used as a stain for the prenyl derivatives). The eluate is frozen in liquid nitrogen and lyophilized overnight.

The pyrophosphate is dissolved in D_2O (10 mg/0.5 mL solvent) and subjected to ^{1}H- and ^{31}P-NMR.

7.4.3
Farnesylation on an Analytical Scale

1. All flasks must be labeled before the experiment is conducted; all devices should be ready and prepared.
2. Calculate the required volumes: 150 µg FTase, 14 nmol H-Ras and 28 nmol isoprenoid pyrophosphate are pipetted together into a microcentrifuge tube, then made up to 100 µL with farnesylation buffer (30 mM Tris/HCl*, pH 7.8, 20 mM KCl, 1 mM $MgCl_2$, 10 µM $ZnCl_2$*, 0.5 mM DTE*) and mixed carefully.
3. Incubate for 1 h at 30 °C.
4. Add Triton-X114 solution (11% (w/v) solution in 30 mM Tris/HCl, pH 7.4, 100 mM NaCl, total volume after addition: 200 µL).
5. Cool for 10 min on ice, shake once in a while.
6. Warm the microcentrifuge tube to 37 °C for 2 min; phase separation takes place.
7. To drive the phase separation to completion, centrifuge the tube for 2 min at 13,000 r.p.m. at ambient temperature.

8. The upper (aqueous) phase is carefully removed with a pipette and transferred to a new microcentrifuge tube.
9. Add 200 μL H$_2$O to the Triton phase (re-extraction of the unconverted Ras protein from the Triton phase).
10. Repeat steps 5–8.
11. Combine the aqueous phases (approx. 300 μL)
12. Add 100 μL H$_2$O to the Triton phase, cool on ice.
13. Add absolute EtOH* to the Triton solution in a 1:1 ratio (v/v); mix and cool on ice; the proteins should precipitate within 30 min.
14. To drive the precipitation to completion, centrifuge the tube for 10 min at 13,000 r.p.m. at 4 °C.
15. Decant the supernatant; the pellet (if visible) should not be destroyed.
16. Redissolve the sediment in 40 μL low-salt-buffer (10 mM Tris/HCl pH 7.4)

All protein-containing samples can be stored overnight at –20 °C.

7.4.4
Farnesylation on a Preparative Scale

Alternatively, prenylation can be carried out on a preparative scale. Scaling up the farnesylation reaction offers the advantage of obtaining native (biologically active) and lipid-modified Ras protein which can be used for further biochemical investigations. Here, removal of Triton-X114 after phase separation is achieved by ion exchange chromatography under non-denaturing conditions. However, the procedure is more time consuming than the *in-vitro* prenylation on an analytical scale.

Procedures
1. All flasks must be labeled before the experiment is conducted; all devices should be ready and prepared.
2. Calculate the required volumes: 2.5 mg FTase, 500 nmol H-Ras wt. fl. and 1 μmol isoprenoid pyrophosphate are pipetted together in a 15-mL centrifuge tube, then made up to 5 mL with farnesylation buffer (30 mM Tris/HCl, pH = 7.8, 20 mM KCl, 1 mM MgCl$_2$, 10 μM ZnCl$_2$, 2 mM DTE) and carefully mixed.
3. Incubate for 3 h at 30 °C. Add 1 μmol farnesyl pyrophosphate after 30, 60 and 90 min of incubation. (Use the incubation period for preparing the column buffer (20 mM Tris/HCl, pH = 7.4, 5 mM MgCl$_2$, 2 mM DTE) and the SDS-PAGE gel)
4. Add 1 mL Triton-X114 solution (11% (w/v) Triton-X114 in 30 mM Tris/HCl, pH = 7.4, 100 mM NaCl).
5. Cool on ice for 10 min, shake once in a while.
6. Incubate for 4 min at 37 °C; phase separation takes place.
7. To drive the phase separation to completion, centrifuge the tube for 4 min at 30 °C and 5000 r.p.m. in a centrifuge with a swing bucket rotor (Eppendorf Centrifuge 5804R or equivalent).

8. The upper (aqueous) phase is carefully removed with a pipette and transferred to a new centrifuge tube.

9. Place the Triton phase on ice and repeat steps 4–8 twice with the aqueous phase (further extraction of lipidated Ras from the aqueous phase).

10. Combine all aqueous phases and save about 100 μL for the SDS-PAGE analysis later on. Combine the three Triton phases (approx. 3 mL) and make up to 10 mL with column buffer.

11. Repeat steps 5–10 with the Triton phase of step 10 in order to re-extract unconverted Ras protein.

12. Dilute the Triton phase 1:10 with column buffer and place the solution into an FPLC loop.

13. Triton-X114 is removed by ion exchange chromatography carried out overnight at 4 °C. For this a compact, automated liquid chromatography system (e. g. ÄK-TA™ prime, Amersham Pharmacia) equipped with a DEAE-Sepharose column is recommended. While the protein binds to the positively-charged column matrix the non-ionic detergent passes through. After removal of the detergent protein elution is achieved by increasing the salt concentration (0–1 M NaCl) of the elution buffer, i.e. by weakening the electrostatic interactions between the Ras protein and charged groups on the column matrix.

14. Pool and concentrate fractions containing the Ras protein using centrifugal ultrafiltration devices with a molecular weight cut-off lower than the molecular weight of the proteins (e. g. Vivaspin 6 mL Concentrator, 10.000 MWCO PES, Vivascience).

15. After determination of the protein concentration, SDS-PAGE-analysis and MALDI-TOF-MS freeze the lipid-modified Ras protein in liquid nitrogen and store at −20 °C.

7.4.5
Determination of Protein Concentration

After *in-vitro* prenylation the following samples are available:

• Aqueous phase of farnesylation experiment: contains unconverted Ras protein, FTase, *E. Coli* proteins, re-extracted proteins from the Triton phase.
• Triton phase after detergent removal: contains farnesylated Ras protein.

The protein concentration is determined via the Bradford assay. BSA (bovine serum albumin) is used as the standard.

The Bradford assay [16] is based on the wavelength shift of Coomassie Brilliant blue G250 dye from $\lambda_{max}=465$ nm to 595 nm when complexed with a protein (for additional information see Chapter 10). To set up a calibration curve 5, 10, 15, 20 and 25 μL of a BSA standard solution (0.4 g/L) are pipetted to 1 mL of a Bradford solution (BioRad, Munich). After mixing well, the extinction is measured at 595 nm compared to water. The absorbance is plotted versus protein concentration which should result in a linear plot.

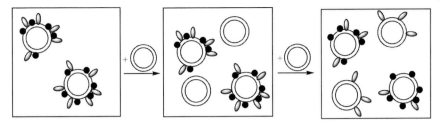

$\mathcal{Q}$ = fluorescent labelled lipidated peptids

● = quencher

addition of empty vesicles;
exchange of lipidated peptids:
increase of fluorescence

Fig. 8.2 Determination of half-time of intervesicle transfer.

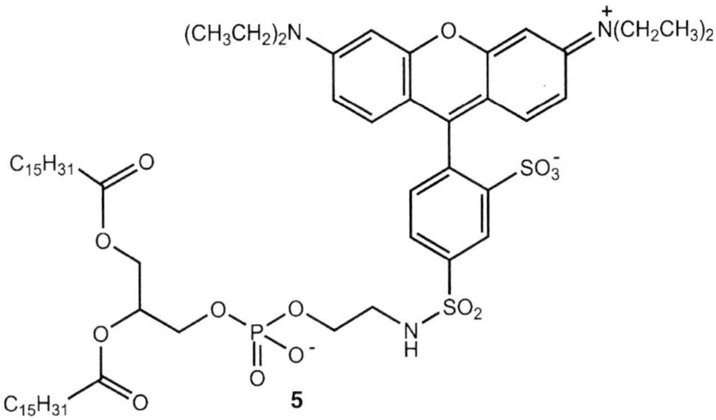

Scheme 8.7 Structure of the fluorescence quencher Lissamine[TM] rhodamine B (**5**).

By plotting the increase in fluorescence intensity versus time, the transfer constant k can be calculated from the slope of the graph (linear regression). Assuming a first order reaction the half-life period can be calculated from k (Eq. 8.2).

$$k = \frac{\mathrm{d}F(t)}{\mathrm{d}t} \qquad t_{1/2} = \frac{\ln 2}{k} \qquad\qquad (8.2)$$

k transfer constant
$F(t)$ fluorescence intensity
$t_{1/2}$ half-life period

In a manner similar to other membrane constituents, proteins and peptides can cross the membrane by transversal diffusion. This so-called flip-flop mechanism should be taken into account because only the peptides from outside the membrane can be transferred to "empty" acceptor vesicles. Peptides bound to the inner surface of the membrane however, will have to cross it first via the flip-flop mechanism.

In an experiment to monitor this process, sodium dithionite is used to reduce the nitro function of the NBD fluorescence marker to an amino function, thus destroying the fluorophore.

After adding the dithionite to the vesicles containing the fluorescent-labeled peptide a rapid decrease in fluorescence can be observed; this is caused by an instantaneous reduction of NBD on the outside of the membrane. Peptides on the inner surface have to cross the membrane first before their fluorophores can be reduced and this results in a much slower decrease in the remaining fluorescence [4]. From this flat part of the resulting graph (fluorescence intensity vs. time) the flip-flop rate can be determined, assuming a first order reaction.

8.3.4
Basic Concepts of Fluorescence and Fluorescence Markers

Definition of fluorescence: if a compound radiates energy within 10^{-10} to 10^{-7} s after excitation (by visible/ultraviolet light, X-rays or electrons) by the same (resonance fluorescence) or a higher wavelength, this radiation is called fluorescence.

Excitation follows the Franck–Condon Principle assuming that electronic transitions are too fast for the nucleus–nucleus distance to readjust (vertical transition). Therefore transition to a higher vibrational state, i.e. not to the vibrational ground state of the first excited electronic state, is common. From this energy level vibrational relaxation to the lowest vibrational state of the first excited electronic state can occur and from there relaxation to the electronic ground state during which the remaining "excess" of energy is radiated as photon (Fig. 8.3). Because of the above-mentioned non-radiating relaxation steps, the amount of energy radiated is usually smaller than that which is absorbed (compare Stokes Rule).

If intersystem crossing-over occurs, a transition from a singlet to a triplet state takes place. This is necessary for phosphorescence. Triplet states show extended half-life periods (approximately 10^{-3} s) because S-T or T-S transitions are spin prohibited; a time delay in radiating the photon results and this kind of radiation is known as phosphorescence.

Fluorescence markers are used for fluorescence spectroscopy of proteins and peptides. These markers can be introduced to a molecule at the *C*- or *N*-terminus or via a side-chain of an amino acid (Scheme 8.8).

8.4
Experimental Procedures

8.4.1
Preparations

The peptide NBD-Aca-Thr-Ile-Cys(Pal)-Ile can be obtained via standard solution phase peptide synthesis as described previously in the literature [5].

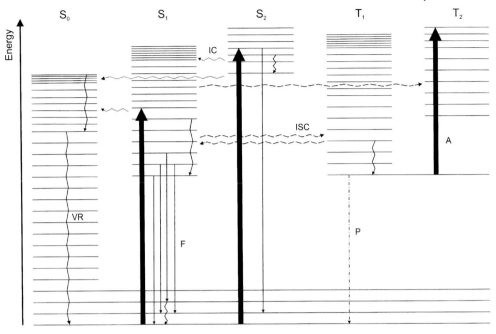

Fig. 8.3   Jablonski diagram. A, absorbtion (➡); F, fluorescence (→); P, phosphorescence (······▶)
ISC, inter-system-crossing ∿∿▶); IC, internal conversion ᠁᠁▶ ; VR, vibrational relaxation
(〰〰▶).

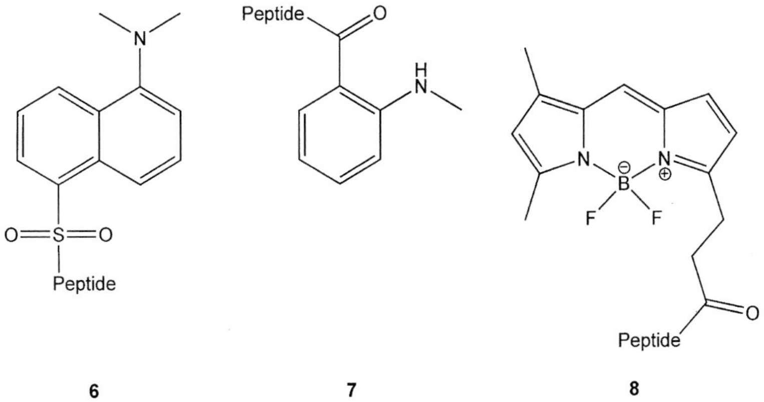

6   **7**   **8**

Scheme 8.8   Structures of fluorescence markers Dansyl (**6**), Mant (methlyeneanthranilic
acid) (**7**) and Bodypy®FL (**8**).

The following solutions will need to be prepared:
- Vesicle buffer (10 mM HEPES, 150 mM KCl, 8 mM NaCl, 1 mM $MgCl_2$, pH = 7)
- 100 μM NBD-Aca-Thr-Ile-Cys(Pal)-Ile-OH in chloroform/methanol 1 : 4
- 200 μM *N*-(Lissamin-rhodamin-sulfonyl)phosphatidyl-ethanolamin in methanol
- 20 mM 1-Palmitoyl-2-oleyl-phosphatylcholin in methanol
- 1 M $Na_2S_2O_4$ in 1 M Tris buffer (pH = 10)

8.4.2
Determination of the Partition Coefficient K_p

8.4.2.1 Preparation of the Peptide–Vesicle Solution
1-Palmitoyl-2-oleyl-phosphatylcholine solution (170 μL, 1 mM, dilution of the 20 mM solution with vesicle buffer) is added to 30 μL of peptide solution (100 μM; $CHCl_3$/MeOH 1 : 4). The vesicles are then prepared by repeated freezing and melting steps. The solvent is removed by evaporation in a centrifuge and the resulting vesicles are dissolved in 200 μL vesicle buffer.

8.4.2.2 Preparation of Solutions of Empty Vesicles of Different Concentrations
Different concentrations of empty vesicle in vesicle buffer are prepared by repeated freezing and melting steps, removal of the solvent and dissolution of the resulting vesicles in vesicle buffer:
- 2 mM: 100 μL of 20 mM 1-Palmitoyl-2-oleyl-phosphatylcholine solution is added to 900 μL vesicle buffer. After removal of the solvent 1000 μL vesicle buffer is added.
- 5 mM: 250 μL of 20 mM 1-Palmitoyl-2-oleyl-phosphatylcholine solution is added to 750 μL vesicle buffer. After removal of the solvent 1000 μL vesicle buffer is added.
- 10 mM: 600 μL of 20 mM 1-Palmitoyl-2-oleyl-phosphatylcholine solution is added to 600 μL vesicle buffer. After removal of the solvent 1200 μL vesicle buffer is added.

8.4.2.3 Measurements by Fluorescence Spectroscopy
The measurements are carried out at an excitation wavelength of 470 nm and an emission wavelength of 540 nm. The cuvette is filled with 2 mL vesicle buffer to which 10 μL of the peptide–vesicle solution is added. F_0 is determined while the solution is being stirred. The various concentrations of empty vesicle solution (as shown in Tab. 8.1) are then added to the solution. F_1 can be determined after 15 min.

8.4.2.4 Interpretation of the Measurements
After successful excitation of the peptides at 470 nm the emitted fluorescence can be measured at 540 nm. The basic principle is that the fluorescence depends on the surrounding matter. The fluorescence intensity (F) of the NBD-labeled peptides is very low in buffer whereas it increases to a maximum value (F_{max}) if the peptide is bound to the membrane. The hyperbolic change in fluorescence intensity, which depends on the concentration of empty vesicles, can be calculated using Eq. 8.3.

Tab. 8.1 Calculation of the coefficient of distribution K_p

Sample no.	Volume added (µL)	Vesicle concentration (mM)	Vesicle amount (mmol/10^{-4})	F_0	F_1	F_1–F_0
1	100	2.0	2			
2	300	2.0	6			
3	400	2.0	8			
4	100	5.0	5			
5	300	5.0	15			
6	400	5.0	20			
7	100	10.0	10			
8	300	10.0	30			
9	400	10.0	40			

$$F = F_0 + (F_{max} - F_0)\frac{K_p^{eff}[L]_{eff}}{1 + K_p^{eff}[L]_{eff}} \tag{8.3}$$

F fluorescence intensity
F_{max} maximum of fluorescence
F_0 initial fluorescence intensity
K_p^{eff} effective partition coefficient given by the concentration of surface-exposed lipids
$[L]_{eff}$ effective concentration of empty vesicles

The distribution of lipidated peptides between aqueous solution and vesicles is shown by Eq. 8.4. With regard to the resulting effective partition coefficient K_p^{eff} the coefficient of distribution K_p of the lipidated peptides can be determined by Eq. 8.5 and 8.6.

$$K_d^{eff} = \frac{1}{K_p^{eff}} \tag{8.4}$$

K_d^{eff} effective dissociation constant that represents the concentration of exposed lipids at which the peptide is 50% partioned into the lipid bilayer

$$K_p = 55.5\,\frac{mol/l}{K_d^{eff}} \tag{8.5}$$

$$K_p = \frac{X_{(peptide)_{vesicle}}}{X_{(peptide)_{aq}}} \tag{8.6}$$

$X_{(peptide)_{vesicle}}$ mol fraction of the peptide on the vesicles
$X_{(peptide)_{aq}}$ mol fraction of the peptide in the aquatic solution

8.4.3
Vesicle Transfer

8.4.3.1 **Synthesis of Vesicles Containing Fluorescence Quencher and Lipidated Peptides**
NBD-Aca-Thr-Ile-Cys(Pal)-Ile-OH solution (15 μL, 100 μM in chloroform/methanol 1:4) and 1-palmitoyl-2-oleyl-phosphatylcholine solution (7.5 μL, 20 mM in methanol) are added to a solution of N-(lissamin-rhodamin-sulfonyl)phosphatidyletha-nolamine (15 μL, 200 μM in methanol), after which the vesicles are generated by repeated freezing and melting steps. The solvent is removed by evaporation in a centrifuge and the resulting vesicles are dissolved in 37.5 μL vesicle buffer.

8.4.3.2 **Preparation of Empty Vesicles**
1-Palmitoyl-2-oleyl-phosphatylcholine solution (30 μL, 20 mM in methanol) is added to 150 μL vesicle buffer, after which the vesicles are obtained by repeated freezing and melting steps. The solvent is removed by evaporation in a centrifuge and the resulting vesicles are dissolved in 180 μL vesicle buffer.

8.4.3.3 **Measurements**
Of the solution containing the vesicles with fluorescence quencher and lipidated peptides 12.2 μL are added to 2400 μL of vesicle buffer and after 30 min (constant baseline) 24 μL (an eight-fold excess) of empty vesicles are added to the solution. The fluorophore is excited at 468 nm and the fluorescence intensity is detected at a wavelength of 535 nm. The measurement is completed after 90 scans (8 h).

8.4.3.4 **Interpretation of the Measurements**
The resulting graph (fluorescence intensity versus time) shows an increase in fluorescence intensity due to the transfer of peptides to the empty vesicles where the fluorescence is not quenched. The resulting initial slope of the graph is determined and Eq. 8.7 can be used to calculate the transfer constant k of the exchange.

$$k = \frac{(dF(t)/dt)_{t=0}}{\Delta F_{max}} \tag{8.7}$$

ΔF_{max} highest possible change of fluorescence
k transfer constant of the exchange

With regard to the transfer constant k, the half-time $t_{1/2}$ can be obtained assuming a first order reaction.

8.4.4
Determination of the Flip-Flop Exchange

8.4.4.1 Measurements
An aliquot of NBD-Aca-Thr-Ile-Cys(Pal)-Ile-OH solution (45 μL, 100 μM in chloroform/methanol 1 : 4) is added to 255 μL of 1 mM 1-palmitoyl-2-oleyl-phosphatylcholine solution (dilution with vesicle buffer). This 17 mM peptide–vesicle solution is obtained by repeated freezing and melting steps, removal of the solvent and dissolution of the resulting vesicles in 300 μL vesicle buffer. Then 2 μL of this 17 mM peptide–vesicle solution are added to 1998 μL of vesicle buffer in a cuvette, after which 20 μL of 1 M $Na_2S_2O_4$ solution in 1 M Tris buffer (pH = 10) is added to the solution. The fluorescence intensity is measured at an excitation wavelength of 470 nm ($\lambda_{excitation}$ = 470 nm; $\lambda_{emission}$ = 540 nm) for 1 h.

8.4.4.2 Interpretation of the Measurements
With regard to the determination of the kinetic parameters of the intervesicle transfer the flip-flop exchange has to be considered. Therefore the transbilayer diffusion has to be determined. After adding the dithionit to the vesicles containing the fluorescent-labeled peptides a rapid decrease in fluorescence can be observed due to the instantaneous reduction of the nitro group of the NBD fluorescence marker. The remaining fluorescence should decrease much more slowly as a result of the peptides having to cross the membrane before the reduction can take place on the surface. From this shallower part of the resulting graph (fluorescence intensity versus time) the flip-flop rate can be determined, assuming a first order reaction.

8.5
Bibliography

8.5.1
Textbooks

VOET, D., VOET, J.G. *Biochemistry*, **1995** John Wiley & Sons, New York, (see relevant chapters).

ATKINS, P. *Physical Chemistry*, W.H. Freeman and Company, New York, **1999**, pp. 503–507.

8.5.2
Special Literature

1 (a) SILVIUS, J.R., ZUCKERMANN, M.J. *Biochemistry* **1993**, *32*, 3153–3161. (b) SILVIUS, R.J., HEUREUX, F. *Biochemistry* **1994**, *33*, 3014–3022. (c) SCHRÖDER, H., LEVENTIS, R., REX, S., SCHELHAAS, M., NÄGELE, E., WALDMANN, H., SILVIUS, J. R. *Biochemistry* **1997**, *36*, 13102–13109.

2 (a) WIESMÜLLER, L., WITTINGHOFER, F. *Cell Signal* **1994**, *6*(3), 247–267. (b) HINTERDING, K., ALONSO DIAZ, D., WALDMANN, H. *Angew. Chem.* **1998**, *110*, 716–780. (c) ZHANG, F.L., CAESEY, P.J. *Ann. Rev. Biochem.* **1996**, *65*, 241–269.

3 Nichols, J. W., Pagano, R. E. *Biochemistry* **1982**, *21*, 1720–1726.

4 (a) HOMAN, R., POWNALL, H. J. *Biochim Biophys Acta* **1988**, *938*, 155–166. (b) MCINTYRE, J. C., SLEIGHT, R. G. *Biochemistry* **1991**, *30*, 11819–11827.

5 EISELE, F., KUHLMANN, J., WALDMANN, H. *Angew. Chem. Int. Ed. Engl.* **2001**, *40*, 369–373.

10

Proteome Analysis: Identification of Proteins Isolated from Yeast

Heinz Prinz, Dörte Goehrke, Petra Janning, and Kerstin Reinecke

10.1
Abstract

Proteins are the main components of living cells, and the ensemble of its proteins characterizes each cell. This ensemble is called the "proteome". In contrast to the genome, which is the same for all cells of a given organism, the proteome varies with the different cell types and with the metabolism of a particular cell. The change of a proteome upon the addition of a drug will provide information regarding its intracellular mode of action.

All strategies employed in proteome research today are based on mass spectrometry in conjunction with the separation of either proteins or peptides (for reviews, see [1, 2]). The classical strategy is as follows: the proteins of a given cell extract are first separated by 2-D gel electrophoresis, the separated proteins are then isolated in small gel pieces. They are then digested enzymatically within their gel matrices. The resulting peptides are eluted from the gel and characterized by mass spectrometry. In most cases, the ensemble of the proteolytic peptide masses ("peptide mass fingerprint") is sufficient for the identification of the isolated protein. Additional sequence information obtained from fragmentation patterns of single peptides (controlled fragmentation inside the mass spectrometer, "MS/MS peptide sequence tag") gives positive proof on the identity of that protein. MS/MS spectra form the basis of automated HPLC-MS proteomics approaches, where a mix of proteins is subjected to an enzymatic digest. The resulting large number of peptides cannot be analyzed by their masses alone. Instead, the MS/MS fragmentation pattern is used for the identification of peptides and their parent proteins. Regardless of the strategy employed, identification is achieved by means of a databank search, which of course, is only possible for known sequences.

10.2
Learning Targets

- Lysis of yeast cells
- Handling of proteins under denaturing conditions

Chemical Biology: A Practical Course.
Edited by Herbert Waldmann and Petra Janning
Copyright © 2004 WILEY-VCH Verlag GmbH & Co. KGaA, Weinheim
ISBN: 3-527-30778-8

- Isoelectric focusing within an immobilized pH gradient
- Reduction of disulfide bonds
- Diffusion of proteins in agarose
- Preparation of polyacrylamide gels
- Denaturation in SDS
- Electrophoresis in cross-linked polyacrylamid gels
- Comparison and quantification of 2-D gels
- Enzymatic digestion of proteins confined in polyacrylamide gel pieces
- Pipetting of small quantities
- MALDI and ESI mass spectrometry of peptides
- Enzymatic digestion of purified proteins
- HPLC-MS electrospray mass spectrometry
- Protein and peptide databanks on the internet
- "Peptide finger print" and "Sequence tag" data bank search

10.3
Theoretical Background

10.3.1
Introduction

Proteome analysis is a developing field in its own right and is also an area of re-search in which the "new economy" biomedical industry has participated with im-pressive enthusiasm. A series of recent reviews can be recommended [1–3]. Since proteome analysis is based on MALDI and ESI mass spectrometry, the instrumen-tation and basic principles will be explained here together with other essential techniques mandatory for the practical tasks.

10.3.2
MALDI Mass Spectrometry

MALDI = **M**atrix **A**ssisted **L**aser **D**esorption **I**onization mass spectrometry has emerged as an effective bioanalytical tool [4]. It uses a pulsed UV laser beam to desorb and ionize co-crystallized sample and matrix from a metal surface.

The "matrix" (e.g. a-cyano-4-hydroxycinnamic acid) absorbs the UV light and is added to the sample in large excess. The absorbed energy causes explosion-like de-sorption of matrix and sample and generates a transient plasma containing ma-trix neutrals (M), matrix ions $(M+H)^+$, $(M-H)^-$, and sample neutrals (A), initially. Within that plasma, numerous collisions of matrix and sample lead to sample molecules which predominantly have only a single proton (Fig. 10.1). The gener-ated ions are accelerated in a strong electric field (Fig. 10.2) which is often applied after the laser pulse ("delayed extraction"). In most MALDI mass spectrometers the analysis is based on the Time-Of-Flight (TOF). Since all molecular ions are ac-celerated with the same energy, small ions have a greater velocity than larger

$$(M+H)^+ + A \rightarrow M + (A+H)^+$$
$$(M-H)^- + A \rightarrow M + (A-H)^-$$

Fig. 10.1 Sample molecules A are ionized by gas phase proton transfer from matrix ions.

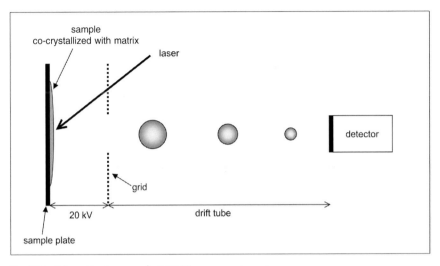

Fig. 10.2 Schematic diagram of a MALDI mass spectrometer.

ones. The time between the application of the electric field and the registration of the signal on the detector can be determined with great accuracy.

However, the ionization energy leads to excitation of the sample molecules and to spontaneous decay of the sample. When samples decay after ionization (PSD=Post Source Decay), the resulting fragments will reach the detector at the same time as the non-fragmented sample, because all will have the same velocity. The peak detected which consists of a cloud of fragments, will be quite broad. Sharp peaks can only be obtained with an ion reflector, an ion optical device which can focus non-fragmented parent masses, and which can also be tuned to study the fragmentation process itself.

The time between laser pulse (or the application of the electric field) and the appearance of a signal on the detector depends not only on the geometry and electric fields of the spectrometer, but also on the geometry of the matrix holder and on the nature of the matrix. Therefore, the experimental error between measurements is relatively large, although the peaks are narrow. This obstacle is corrected when standards are used. External standards (samples of similar nature placed inside the same matrix at a spot near the actual sample) or internal standards (compounds of known mass mixed with the sample) can be used.

10.3.3
Electrospray Mass Spectrometry using Ion Trap Analyzers

10.3.3.1 **Principle**
Electrospray [5] is a method by which a solution is dispersed in the presence of an electric field creating a fine spray of highly charged droplets (Fig. 10.3). Dry gas or heat is applied to the droplets causing the solvent to evaporate. The spray is transferred into the vacuum of a mass spectrometer. The volume of the droplets is further decreased and thereby the surface charge and curvature is increased.

Sample ions are expelled and can be analyzed in the mass spectrometer using an ion trap, a quadrupole or a conventional sector field mass spectrometer. Note that no ionization energy is applied. Ions which are transferred to the vacuum by electrospray are not in an excited state and will not decay spontaneously inside the spectrometer. The machine, which will be used during our practical course, is a Finnigan LCQ which is shown schematically in Fig. 10.4.

The droplets, once they are formed at the tip of an HPLC outlet, are completely evaporated by means of a heated capillary. A tube lens and skimmer filters out the ions of interest (typically positive ions). An ion optic array of quadrupole, lens and octopole focuses the ion beam into the ion trap. A broad radio frequency range of low amplitude is applied to that trap, so that a broad range of ions is trapped therein (typically m/z between 150 and 2000). Once the trap is "full"

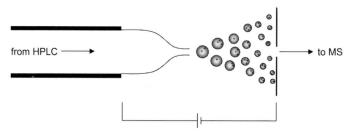

Fig. 10.3 Schematic diagram of ionization in ESI.

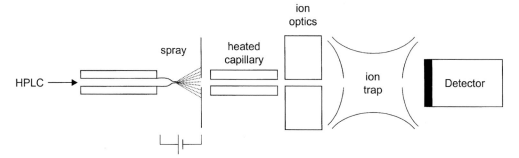

Fig. 10.4 Schematic diagram of an electrospray mass spectrometer.

2. *Preparation of the re-swelling tray*: Slide the protective lid completely off the tray and level the tray by turning the leveling feet until the bubble in the spirit level is centered. Ensure that the tray is clean and dry.

3. *Application of the rehydration solution*: Pipette 400 µL of the supernatant from the cell lysate into each slot. Deliver the solution slowly into the center of the slot. Remove any large bubbles. Important: To ensure complete sample uptake, do not apply excess rehydration solution.

4. *Positioning of the IPG strip*: Remove the protective cover from the IPG strip and position it with the gel side down and the pointed end against the sloped end of the slot. Lower the IPG strip onto the solution. To help coat the entire IPG strip, gently lift and lower the strip and slide it back and forth along the surface of the solution. Be careful not to trap bubbles under the IPG strip. Overlay each IPG strip with 1.5 to 3 mL of a paraffin oil (IPG Cover Fluid, Pharmacia, or equivalent) to minimize evaporation and urea crystallization.

5. *Rehydration of the IPG strip*: Slide the lid onto the re-swelling tray and allow the IPG strips to rehydrate at room temperature. A minimum of 10 h is required for rehydration and usually proceeds overnight at room temperature.

6. *Sample application:* When the rehydration cassette has been completely emptied and opened, the strips are transferred to the corresponding strip tray. After placing the IPG strips, humid electrode wicks, electrodes and sample cups in position, the strips and cups are covered with low viscosity paraffin oil. Samples are applied at the cathodic end of the IPG strips in a slow and continuous manner without touching the gels.

7. *Running conditions:* The voltage is increased linearly from 300 to 3500 V over 3 h, followed by three additional hours at 3500 V, whereupon the voltage is increased to 5000 V. A total volt–hour product of 100 kVh is used in an overnight run.

8. *Equilibration of* IPG gel strips: After the first-dimension run, the strips are equilibrated in order to re-solubilize the proteins and to reduce -S-S- bonds. The strips are equilibrated within the strip tray with 100 mL of reduction buffer for 12 min. The -SH groups are subsequently blocked with 100 mL of alkylation buffer for 5 min.

10.4.3.3 Second Dimension: SDS-PAGE

In the second dimension, a vertical gradient slab gel is used with the Laemmli–SDS discontinuous system with two small modifications. (i) Gels are not polymerized in the presence of SDS. This seems to prevent the formation of micelles which contain the acrylamide monomer, thus increasing the homogeneity of pore size and reducing the concentration of non-polymerized monomer in the polyacrylamide. The SDS used in the gel running buffer is sufficient to maintain the necessary negative charge on the proteins. (ii) The combination of the IPG strip and agarose avoids the need for a stacking gel.

1. *Pouring gels for the second dimension:* 27.5 mL separation buffer, 44 mL acrylamid/bis and 38.5 mL water are mixed yielding the acrylamide gel. Just before

pouring the gel a solution of 0.1 g ammonium persulfate in 150 µL water and 50 µL TEMED is added.

2. *Transfer of IPG gel strips:* after equilibration, the IPG gel strips are cut to size; 6 mm are removed from the anodic end and 14 mm from the cathodic end. The second dimension gels are overlaid with a solution containing agarose (1% w/v in running buffer), heated in a microwave and cooled to ca. 80 °C. The IPG gel strips are immediately loaded through the agarose.

3. *Running the second dimension:* the voltage is applied until the front which is made visible by the bromophenol blue, reaches the bottom of the gel. The running conditions are as follows: current, 40 mA/gel (constant); the voltage is non-limiting, but usually requires 100 to 400 V; temperature, 8–12 °C; duration of run, 3–4 h.

4. *Staining the proteins with colloidal Coomassie dye:* when the visible bromophenol blue has eluted from the gel, the voltage is switched off and the gels are carefully transferred into a tray containing staining solution. Note that the fragile gels may be torn when too much force is applied. Therefore the glass plates should be separated very slowly, and the gel that typically sticks to one of the plates should not be touched. Gravity will be sufficient to let the gel slip into the staining solution. If necessary the gel could be wetted using this solution. The gels are then gently shaken in the staining solution at room temperature overnight. They are de-stained in 50% ethanol and 10% acetic acid for several hours. After 1 h The de-staining solution is replaced ca. once per hour. The gel is welded into a plastic foil and then placed on a scanner or photocopy machine in order to record the localization of the protein spots. The spots are isolated and transferred into Eppendorf tubes which are then stored in the freezer.

10.4.4
In-gel Digestion of Proteins Separated by Polyacrylamide Gel Electrophoresis

The following protocol has been adapted from an EMBO course organized by the Protein and Peptide Group EMBL Heidelberg [8].

1. *Excision of protein bands from polyacrylamide gels:* wash the gel slab with water (two changes, 10 min each) and excise spot(s) of interest with a clean scalpel. Cut as close to the protein band as possible to reduce the amount of "background" gel. Cut excised piece into ca. 1×1 mm cubes and transfer them to a 1.5-mL microfuge tube.

2. *Washing the gel pieces:* remove all remaining liquid and add sufficient acetonitrile to cover the gel pieces. After the gel pieces have shrunk (they become white and stick together) remove acetonitrile and rehydrate with 20 µL 0.1 M NH_4HCO_3. After 5 min, add 20 µL of acetonitrile (to give 0.1 M NH_4HCO_3/acetonitrile, 1 : 1). Remove all liquid after 15 min of incubation. Dry down gel particles in a vacuum centrifuge ("speed-vac").

3. *Option of silver-stained gels:* instead of the above rinsing regimen, just shrink the gel pieces with acetonitrile and then reduce/alkylate as described below.

4. *Reduction and alkylation*: swell the gel particles with 20 μL 10 mM dithiothrei-tol/0.1 M NH₄HCO₃ and incubate for 45 min at 56 °C to reduce the protein. Chill tubes to room temperature. Remove excess liquid, shrink gel pieces with 20 μL acetonitrile and replace it quickly with 20 μl 55 mM iodoacetamide in 0.1 M NH₄HCO₃. Incubate for 30 min at room temperature in the dark. Remove iodoacetamide solution, and wash the gel particles with 0.1 M NH₄HCO₃ and acetonitrile as described above (Washing of gel pieces). (Note: all Coomassie should be removed at this time and gel particles should look completely transparent. If large amount of protein is analyzed (more than 5 pmol) residual staining may still be observed. In this case an additional washing cycle with 0.1 M NH₄HCO₃ and acetonitrile should be carried out).

5. *In-gel digestion with trypsin:* add 20 μL acetonitrile, wait 5 min and dry down the gel particles completely in a Speed Vac. Rehydrate gel particles with 20 μl digestion buffer containing 50 mM NH₄HCO₃, 5 mM CaCl₂ and 12.5 ng/μL trypsin (e.g. Boehringer Mannheim, sequencing grade, or Promega, modified, sequencing grade) at 4 °C (ice bucket). Add sufficient digestion buffer to cover the gel pieces. Add more buffer if the volume added initially has been absorbed by the gel pieces. After 45 min remove remaining supernatant and replace it with 5–20 μL of the same buffer – but without trypsin – to keep the gel pieces wet during enzymic cleavage (37 °C, overnight).

6. *Extraction of peptides:* a small (0.3–0.5 μL) aliquot of digestion buffer is taken up for MALDI analysis. If subsequent nano ESI-MS/MS analysis is necessary the gel pieces should be subjected to additional extraction procedures. The peptides are extracted from the gel by addition of a sufficient volume (20 μL) of 25 mM NH₄HCO₃ to cover the gel pieces. Incubate for 15 min and add the same volume of acetonitrile. Incubate for 15 min and recover the supernatant. Repeat the extraction with 5% formic acid (swelling) followed by acetonitrile (shrinking). Pool all the extracts and dry the sample in a vacuum centrifuge.

10.4.5
MALDI Mass Spectrometry

The aliquots of the peptide mixtures are mixed with 4 μL saturated a-cyano-4-hydroxycynamic acid in 0.1% TFA/acetonitrile 2 : 1 of which 1 μL is spotted onto a MALDI plate. Once it has dried, 2 μL ice-cold 0.1% TFA is added on top of the spot and immediately removed either by aspiration or by shaking the plate. This process (washing the plates) reduces the salt concentration and enhances the sensitivity. The MALDI measurements are calibrated by adding a spot containing standards near the samples. These standards are peptides with known masses. As mass standards, the following peptides with different monoisotopic mass to charge ratios $[M+H]^+$ can be used: 904.4681 (Bradykinin), 1296.6853 (Angiotensin I), 1570.6774 (Glu-Fibrino-Pep B), 1296.6853 (Angiotensin I), 2093.0867 (ACTH (1-17)), 2465.1989 (ACTH (18-39)), 3657.9294 (ACTH (7-38)), and 5730.6087 (Insulin (bovine)). This covers the whole range of masses of interest.

10.4.6
Alternative: HPLC ESI-MS/MS

10.4.6.1 **Principle**
One alternative to the conventional 2-D PAGE (in conjunction with the identification of the spots by mass spectrometry) is an on-line method which involves two-dimensional HPLC and on-line electrospray mass spectrometry. For the first dimension, a peptide mix is bound to strong ion exchange columns. The peptides are eluted stepwise with increasing salt concentrations and transferred to a conventional HPLC column. From there, they are eluted with gradients of a volatile organic solvent such as acetonitrile. The flow is transferred directly into a mass spectrometer and analyzed both by its mass and by its MS/MS fragmentation pattern. For the practical course, we will only run the "conventional" reverse phase HPLC coupled to an ESI mass spectrometer.

10.4.6.2 **Protein Quantification**
In order to perform enzyme digestion, the protein content of a given sample must be known. During the practical course the students will compare the results of two popular assays, the Bradford- [9] and the BCA(bicinchoninic)-assay [10]. They should learn that protein assays may be influenced by other reagents in the solution. In this case the Bradford assay is influenced by CHAPS (included in the lysis-buffer), therefore the results obtained from the BCA-assay are more reliable.

Bradford Assay
This assay is based on the specific binding of the dye Coomassie blue (Scheme 10.2) to proteins at arginine, tryptophan, tyrosine, histidine and phenylalanine residues with an eight-fold greater affinity for arginine. It should be noted that since the assay primarily responds to arginine residues, any analysis of an arginine-rich sample will need to include an arginine-rich standard. The dye binds to these residues in the anionic form, which has an absorbance maximum at 595 nm (blue). The free dye in solution is in the cationic form, which has an absorbance maximum at 470 nm (red). The assay is monitored at 595 nm in a spectrophotometer, and thus measures the protein-dye complex.

The assay has to be calibrated against a reference protein. We will use 2, 4, 8, 12, 16 and 20 µg bovine serum albumin (BSA) as standards (pipetted from a stock solution of 1 mg/mL BSA and added to a 20-µL sample volume with water). The stock solution of Bradford reagent consists of: 50% (v/v) H_3PO_4, 25% (v/v) ethanol, 0.05% (w/v) Serva Blue G. The reagent is diluted 1 : 5 with H_2O prior to the experiments. The samples (1, 2, 5 and 10 µL of the cell lysate and the standards) are added to 1 mL of the diluted reagent. The absorption at 595 nm is recorded after 10 min incubation time at room temperature. Data are evaluated graphically by comparison with the standards.

11

Lectins: Determination of the Sugar Specificity of Jacalin by a Sugar–Lectin Binding Assay (SLBA)

Eleni Gourzoulidou and Stefanie Schlummer

11.1
Abstract

Lectins are proteins or glycoproteins which can reversibly bind to specific segments of carbohydrates through hydrophobic interactions. They have been identified in all known organisms and occur in manifold molecular structures. They have multiple functions in different organisms, but always identify and bind specifically to mono- or oligosaccharides.

In this experiment Jacalin will be used as an example for the sugar selectivity of lectins. Jacalin is isolated from the seeds of the jackfruit (*Artocarpus integrifolia*) and has been found to selectively bind *a*-O-glycosides, which are derived from galactose (Gal) or *N*-acetyl-galactosamine (GalNAc) segments.

The selectivity of Jacalin will be determined in a sugar–lectin binding assay (SLBA), a modified version of the enzyme-linked immunosorbent assay (ELISA). First, a biotin–galactose conjugate will be synthesized, which will then be used in a SLBA to bind to the immobilized lectin in conjunction with various competing carbohydrates. The bound conjugate will then be detected by complexation to streptavidin–horseradish peroxidase.

11.2
Learning Targets

- Lectin specificity
- Biotinylation of carbohydrates
- Biotin–streptavidin complex
- Principle and performance of an ELISA-related binding assay
- Evaluation by microtiterplate reader

Chemical Biology: A Practical Course.
Edited by Herbert Waldmann and Petra Janning
Copyright © 2004 WILEY-VCH Verlag GmbH & Co. KGaA, Weinheim
ISBN: 3-527-30778-8

11.3
Theoretical Background

11.3.1
Lectins

Lectins [1–3] were first discovered by Stillmark in 1888. He found that the extract of castor bean (*Ricinus communis*) seeds could agglutinate erythrocytes. Later, in the 1940s it was found that agglutinins could select types of cells based on their blood group activities. The specificity of lectins against various cell types is based on the different glycosylation patterns on their surfaces. Lectins were originally detected and isolated from plants, occurring in high quantities in the seeds. In the meantime, lectins were found in all types of organisms including viruses, bacteria, protozoa, fungi, plants, invertebrates and vertebrates, in various molecular forms. Although the name lectin (from the Latin: *lectus = selected*) was originally used to define agglutinins which could discriminate different types of red blood cells, today the term is used generally and includes sugar-binding proteins from various different sources. Lectins are proteins or glycoproteins of non-immune origin with one or more binding sites per subunit, which can reversibly bind to specific sugar segments through hydrogen bonds and van der Waals interactions. The designation of lectins in the literature is based either on the scientific name of the species from which they were isolated or in terms of their monosaccharide specificity.

Lectins are not only important for the agglutination of erythrocytes. Due to the fact that the cell surfaces are covered with glycoproteins, they also play an important role in all cell–cell adhesion processes such as sperm–egg interactions, inflammatory responses or cell–virus and cell–bacteria interactions. These carbohydrates occur as oligosaccharides bound to membrane proteins (glycoproteins) and lipids (glycolipids) and as integral membrane proteoglycans.

They can be isolated in high yield and purity by chromatographic techniques such as affinity chromatography. Although lectins are ubiquitous in nature, the biological processes in which they are involved are not completely understood. Their function varies from organism to organism, but they all work by specifically recognizing mono- and oligosaccharides.

Jacalin, one of two lectins in Jackfruit (*Artocarpus integrifolia*) seeds, was first isolated in 1979 [4]. Jackfruit (Fig. 11.1) is an edible, large, up to 90 cm long, pear- or barrel-shaped fruit that grows abundantly in several tropical countries. Jacalin consists of four subunits, which are associated through non-covalent interaction. Two of the subunits are of approximately 10 kDa and two of 16 kDa (Scheme 11.1) [5].

Because of its unique property of binding human antibody IgA, crude Jacalin is purified by affinity chromatography on immobilized IgA or IgA1, the bound protein being eluted with galactose [6]. The fact that Jacalin selectively binds human secretory IgA with a specificity for the IgA1 subclass, was discovered at the end of 1970s and has been exploited for the isolation of IgA1. Previously the purification of human IgA had been very tedious and affinity chromatography with agarose-bound Jacalin became a convenient procedure for its isolation [6].

Fig. 11.1 Jackfruit.

Jacalin predominantly binds α-O-glycosides of Gal or GalNac moieties [7]. It is highly specific for the Thomsen–Friedenreich antigen (Galβ1-3GalNAc), which is a T-antigenic disaccharide associated with tumors and generally specific for O-linked glycoproteins [6].

Recently an application for Jacalin was found in AIDS research due to its unique mitogenic properties with regard to CD4$^+$ cells [8]. Jacalin-induced T-cell proliferation can provide information about the deficiency of CD4$^+$ T-cells in HIV-1 infection. Native Jacalin as well as a Jacalin-peptide derivative have been found to prevent HIV-1 infection *in vitro* [9]. However further research is required to elucidate the mechanism of action of Jacalin in preventing HIV-1 infection and its mode of interaction with CD4$^+$ cells as these processes are not well understood at present. Further investigation is also necessary to evaluate whether Jacalin is a potential therapeutic agent in the prevention of HIV-1 infection.

Many plant lectins, including Jacalin, have been used as probes for organizational and structural characterization of glycoconjugates on the cell surface. Studies with lectins have proved invaluable in understanding the molecular mechanisms of cell growth differentiation, malignancy, immune response, membrane structure and organization [10]. In order to utilize lectins as probes in biological assays, it is necessary to determine their properties and specificity for sugars and glycoconjugates.

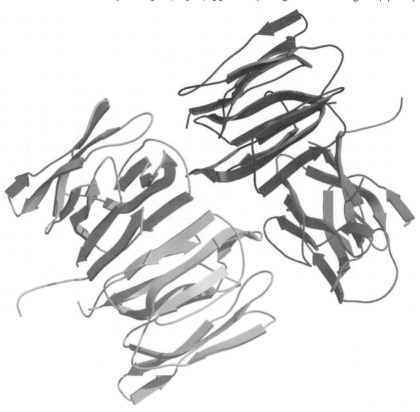

Scheme 11.1   Structure of Jacalin (according to Vijayan *et al.* [5]).

11.3.2
Sugar–Lectin Binding Assay (SLBA)

The selectivity of Jacalin will be determined in a sugar–lectin binding assay (SLBA) [11] by inhibiting binding between Jacalin and D-galactose with a series of different carbohydrates (Scheme 11.2) which act as competing substrates. This assay is a modification of the enzyme-linked immunosorbent assay (ELISA). Instead of the detection of an antigen–antibody complex, the binding between a lectin and a sugar is monitored in a SLBA. Several different variants of ELISAs exist, which differ mainly in the sequence of addition of antibody, analyte and tracer. ELISAs are classified mainly as direct or indirect competitive immunoassays. Due to their high selectivity and sensitivity, ELISAs are used successfully as a standard method in medicinal analytics.

The SLBA is a heterogeneous and competitive assay (Fig. 11.2) and can be used to measure small molecules [11]. First, the lectin Jacalin is immobilized on the bottom of a multi-well plate well and the excess is washed away. To block any remaining protein binding capacity of the plastic well, the plate is incubated with which 150 µL *o*-phenylenediamine (0.35 mg/mL, dissolved in buffer G) is added. The plate is left in the dark at room temperature for 15 min. The reaction is terminated by the addition of 50 µL of 2 M sulfuric acid*.

The color intensity is measured at a wavelength of 492 nm in a microtiter plate reader.

To determine the sugar specificity of Jacalin by competition or inhibition of binding of the BG conjugate, a series of carbohydrates of different concentrations is added to the well together with the BG conjugate.

Concentrations of carbohydrates used: 0.01, 0.05, 0.1, 0.2, 0.5, 1.0, 1.5, and 2.0 mM.

Carbohydrates used: D-Galactose, methyl-*a*-D-galactoside, methyl-*a*-D-galactoside, *N*-acetyl-D-galactosamine and D-glucose.

11.4.4
Calculations

The measured extinctions should be in the linear range according to Lambert–Beers Law. First, all extinctions have to be corrected by the background. The signal caused by the binding of the BG conjugate to Jacalin in the absence of any further sugar is equivalent to 100 % activity of the BG conjugate. By the addition of increasing amounts of a competitive sugar, the signal should decrease due to the competition between BG and the other sugar for the lectin binding sites. Protraction of % inhibition against the competitor concentration yields a saturation curve, from which the concentration of competitor required for half-maximum inhibition (IC_{50}) can be determined (Fig. 11.3).

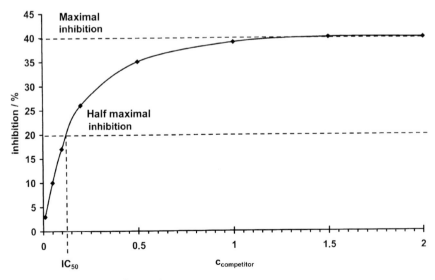

Fig. 11.3 Determination of IC_{50} value.

Tab. 11.2 Literature values for maximum inhibition and IC_{50} of different carbohydrates (Wetprasit *et al.* [11])

Carbohydrates	Maximum inhibition (%)	IC_{50} (mmol/L)
D-Galactose	57.99	0.2
Methyl-α-D-galactoside	21.72	0.035
Methyl-β-D-galactoside	7.86	0.33
N-Acetyl-D-galactosamine	42.38	0.29
D-Glucose	5.93	0.475

For any particular sugar, the lower its IC_{50} value, the higher its binding affinity or specificity for Jacalin. Experiments by Wetprasit *et al.* [11] demonstrated that Jacalin is more specific for methyl-α-D-galactoside ($IC_{50}=0.035$ mM) than for D-galactose ($IC_{50}=0.2$ mM) (Tab. 11.2).

11.5
Bibliography

1 (a) LIS, H., SHARON, N. *Ann. Rev. Biochem.* **1986**, *55*, 35–67. (b) SHARON, N., LIS, H. *Lectins*, Chapmann and Hall, London, **1989**.

2 KENNEDY, J. F., PALVA, P. M. G., CORELLA, M. T. S., CAVALCANTI, M. S. M., COELHO, L. C. B. B. *Carbohydr. Polym.* **1995**, *26*, 219–230.

3 SINGH, R. S., TIWARI, A. K., KENNEDY, J. F. *Crit. Rev. Biotechnol.* **1999**, *19*, 145–178.

4 CHATTERJEE, B., VAITH, P., CHATTERJEE, S., KARDUCK, D., UHLENBRUCK, G. *Int. J. Biochem.* **1979**, *10*(4), 321–327.

5 SANKARANARAYARAN, R., SEKAR, K., BANERJEE, R., SHARMA, V., SUROLIA, A., VIJAYAN, M. *Nature Struct. Biol.* **1996**, *3*, 596–603.

6 KABIR, S. *J. Immunol. Methods* **1998**, *212*, 193–211.

7 WU, A. M., SUGII, S. *Carbohydr. Res.* **1991**, *213*, 127.

8 PINEAU, N., AUCOUTURIER, P., BRUGIER, J. C., PREUDHOMME J. L. *Clin. Exp. Immun.* **1990**, *80*, 420–425.

9 CORBEAU, P., HARAN, M., BINZ, H., DEVAUX, C. *Mol. Immunol.* **1994**, *31*(8), 569–575.

10 MODY, R., JOSHI, S., CHANEY, W. *J. Pharmacol. Toxicol. Methods* **1995**, *33*, 1–10.

11 WETPRASIT, N. W., CHULAVATNATOL, M. *Biochem. Mol. Biol. Int.* **1997**, *42*, 399–408.

Phenylhydrazides can be oxidized to acyldiazenes which are known to fragment upon treatment with nucleophiles to yield carboxylic acid derivatives. Waldmann and coworkers [3] have recently developed a traceless linker for arenes based on this methodology.

13.3.3
The Hydrazide Linker: An Oxidation Labile Traceless Linker

The hydrazide linker [3] serves as a traceless linker that exhibits high stability and allows orthogonal and selective cleavage under mild oxidative conditions. The oxidative cleavage of aryl hydrazides forming carboxylic acid derivatives, dinitrogen, and arenes has been applied in peptide chemistry both in solution and on a solid support to obtain peptide carboxylic acids, amides, and esters (Scheme 13.2).

In general, this cleavage is a two-step process consisting of oxidation of the hydrazide **1** to the activated acyldiazene intermediate **2** and its subsequent nucleophile-induced fragmentation. By using catalytic amounts of Cu(II) salts as the oxidative reagent in the presence of a nucleophile and pyridine (for the complexation of Cu^{II}), the cleavage cannot be stopped at the stage of the active intermediate **2**. In contrast, the use of N-bromosuccinimide in the presence of pyridine offers the opportunity to isolate the activated acyldiazene **2**. Its fragmentation can be initiated by adding a solution of the nucleophile. Furthermore, it has been shown that phenylhydrazides can be cleaved enzymatically using a tyrosinase isolated from mushrooms [4].

Previous work by Semenov *et al.* [5] showed that arylhydrazides are sensitive to oxidative cleavage on a solid support and are thus suitable for the synthesis of peptides (Scheme 13.2, R^1=peptide, aryl=phenyl moiety linked to a solid support). Oxidation of polymeric hydrazide **1** was carried out using copper(II)-sulfate in a mixture of DMF, acetic acid, and pyridine acetate buffer.

The reverse situation i.e. the attachment of phenylhydrazides to an acid-functionalized polymeric support, offers the opportunity to cleave off aryl compounds in a traceless manner by creating an aryl-hydrogen bond (Scheme 13.3).

Since phenylhydrazides are sensitive to oxidative cleavage their use as traceless linkers facilitates the synthesis of a variety of differently substituted aromatic compounds on solid support and their cleavage under extremely mild conditions. Because hydrazides are stable to acids and bases the traceless phenylhydrazide linker is generally applicable to solid phase and combinatorial chemistry.

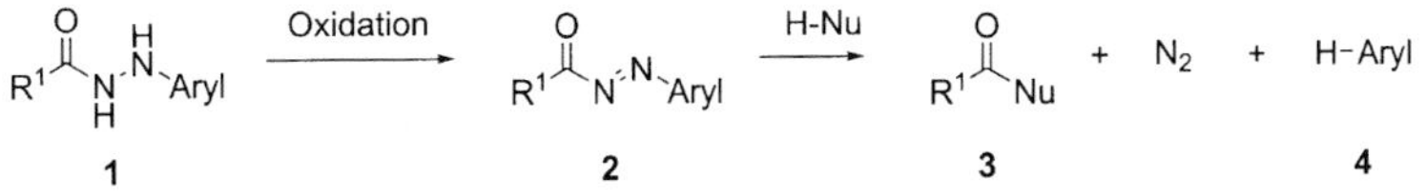

Scheme 13.2 Oxidative cleavage of hydrazides

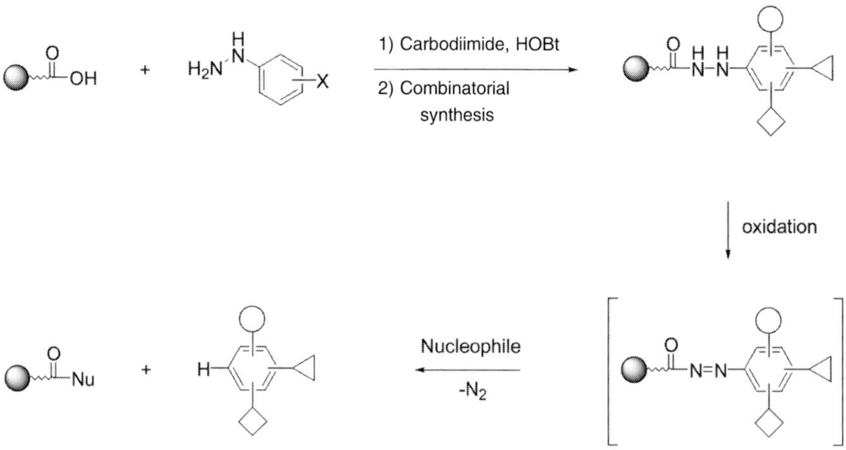

Scheme 13.3 Concept of the traceless phenylhydrazide linker

13.3.4
Acid-functionalized Resins

In order to use hydrazide as a traceless linker group, polymeric supports carrying carboxylic acids on their surface are needed to couple different hydrazines to the solid support before subjecting them to combinatorial derivatization and final traceless oxidative cleavage (Scheme 13.3).

Inspection of the resins obtained after activation of the polymer-bound carboxylate with carbodiimide and subsequent treatment with different phenylhydrazines by means of Fourier transform IR spectroscopy, consistently revealed a strong band at $1705\,cm^{-1}$ which could not be ascribed to the starting material or the product. The commercially available resins **5** employed in these experiments had been obtained from an amino-functionalized carrier by formation of a succinic acid monoamide, i.e. the resin had been functionalized with an amide-derived nitrogen and an activated carboxylic acid at a distance suitable for the formation of a five-membered ring imide. Thus, in order to explain the finding detailed above it was speculated that after reaction of the carboxylic acid **5** with the carbodiimide, there is an intramolecular attack on the activated intermediate by the nitrogen of the amide group to finally give rise to the polymer-bound cyclic imide **6** (Scheme 13.4). Under basic conditions the polymer-bound succinimide **6** was readily hydrolyzed to give acid **5**.

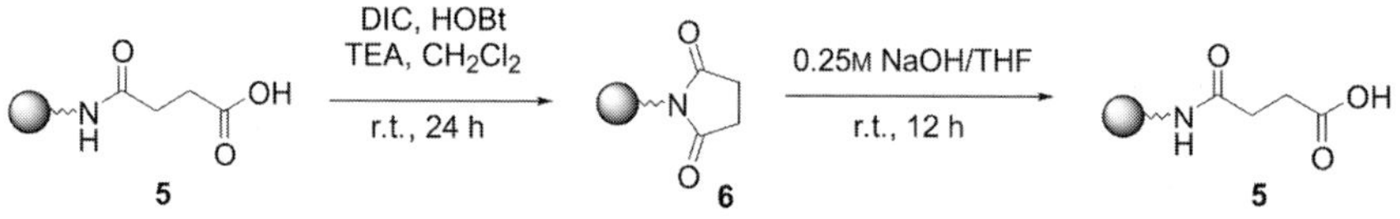

Scheme 13.4 Cyclization of the acid-functionalized polymers **5** to the succinimide **6** under conditions of hydrazide formation

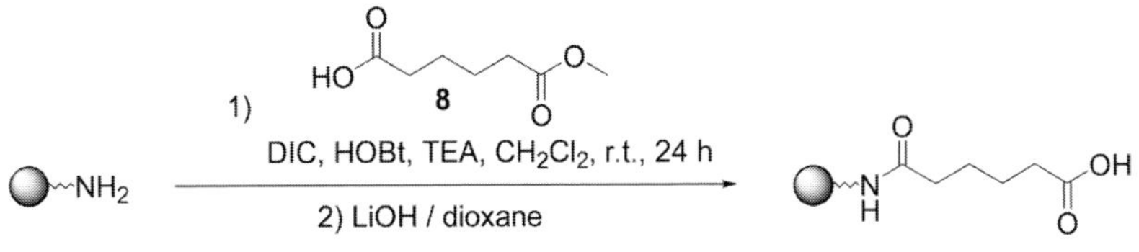

Scheme 13.5 Synthesis of adipic acid-functionalized resin **9**

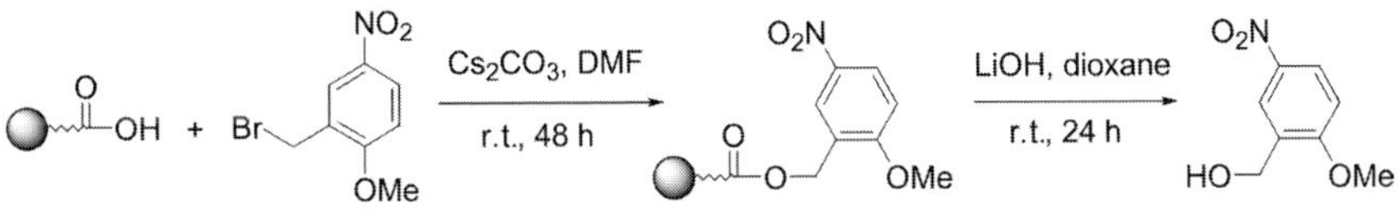

Scheme 13.6 Synthesis of polymer-bound 2-methoxy-5-nitrobenzylester **11** for determination of loading

In attempting to overcome this problem a simple elongation of the carbon chain by two methylene units, i.e. the use of an adipic acid derivative instead of a succinic acid derivative, provided a straightforward solution since the formation of a seven-membered ring is unfavorable. The acid-functionalized resins obtained in this way do not undergo undesirable cyclization upon activation with carbodiimide (Scheme 13.5).

Amino-functionalized polystyrene **7** is reacted with adipic acid methyl ester **8** to yield an adipic acid monoamide **9**, followed by basic saponification of the ester group (Scheme 13.5). The loading level is then determined by nucleophilic esterification of the carboxylic acid groups **9** with 2-methoxy-5-nitrobenzyl bromide **10** in DMF and basic saponification of the resulting polymeric esters (Scheme 13.6).

By means of UV spectroscopy (at 307 nm) loading can be quantified by determining the amount of 2-methoxy-5-nitrobenzyl alcohol **12** released.

13.3.5
Solid-phase Synthesis of Antibiotics

A robust guideline for ensuring that the correct choice of suitable biologically-relevant molecular scaffolds is made, is the principle of identifying "privileged structures", i.e. compound classes that facilitate interactions with various proteins. Biphenyls are found in various pharmacologically-active compounds, e.g. in vitronectin receptor antagonists, angiotensin receptor antagonists, inhibitors of transthyretin-mediated amyloid fibril formation and novel antibacterial agents [3]. In addition, the biphenyl unit has been proposed as a general scaffold for the combinatorial generation of new drug candidates.

Biphenyl derivative **22a** and its morpholine analog **22b** (Scheme 13.7) are synthesized using the traceless hydrazide linker. Compound **22a** is a representa-

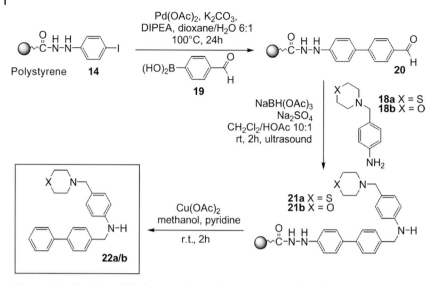

Scheme 13.7 Traceless solid-phase synthesis of antibiotic **22a** and its derivative **22b** employing the traceless hydrazide linker

tive member of a recently discovered new class of antibiotics active against *Mycobacterium tuberculosis* (i.e. the bacteria that causes tuberculosis) and against atypical mycobacteria.

The acid-functionalized resin **9** is used for the attachment of 4-iodophenylhydrazine **13** by activation with *N,N*-diisopropylcarbodiimide (DIC) and 1-hydroxybenzotriazole (HOBt) (Scheme 13.8).

The iodophenylhydrazide **14** is reacted with 4-formylphenylboronic acid **19** to give rise to the polymer-bound biphenylaldehyde **20** (Scheme 13.7). The resulting biphenyl aldehyde **20** is then subjected to reductive amination with thiomorpholine derivative **18a** and the morpholine derivative **18b** to yield the polymer-bound secondary amine **21a** and **21b**. The desired biphenyl antibiotic **22a/b** is cleaved off under oxidative conditions with Cu(OAc)$_2$ in methanol and pyridine (Scheme 13.7). Further purification is carried out using SPE cartridges (solid phase extraction).

Both aniline derivatives **18a/b** are synthesized by two-step procedures starting with the reaction of 4-nitrophenylmethyl bromide **16** and thiomorpholine **15a** respectively morpholine **15b** in THF at room temperature for 2 days to yield 4-(4-nitrophenylmethyl)-thiomorpholine **17a** and 4-(4-(4-nitrophenylmethyl)-morpholine

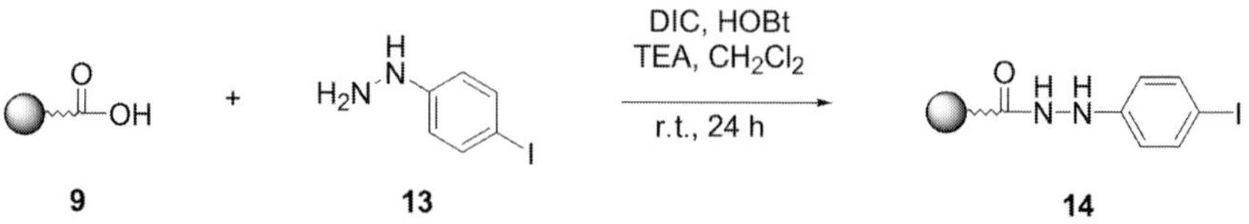

Scheme 13.8 Synthesis of polymer-bound iodophenylhydrazide **14**

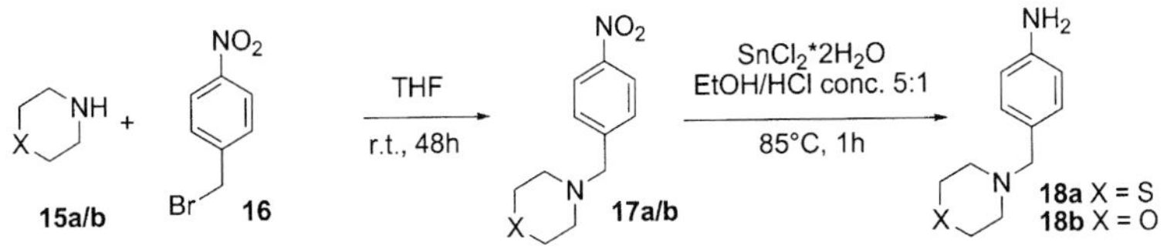

Scheme 13.9 Synthesis of the 4-morpholinomethylaniline derivatives **18 a/b**

17 b, respectively. Following this the aniline derivatives **18 a/b** are obtained by re-duction with stannous dichloride (Scheme 13.9).

13.4
Experimental Procedures

13.4.1
Project A: Synthesis of Functionalized Resin

13.4.1.1   Procedure for the Preparation of Adipic Acid Monomethyl Ester-functionalized Supports

To a suspension of 1 g amino-functionalized polystyrene-resin **7** in methylene chloride (30 mL) *N,N*-diisopropylcarbodiimide* (3 eq.), 1-hydroxybenzotriazole* (3 eq.), triethylamine* (3 eq.) and adipic acid monomethylester **8** (3 eq.) are added and the mixture is shaken for 18 h at room temperature. The mixture is filtered, the resin is washed with methylene chloride, THF*, THF/1 N HCl (1:1), THF, methanol, methylene chloride and cyclohexane (10 mL, two times each) and dried to constant weight *in vacuo*.
Task: IR measurements and assignment of characteristic bands.

13.4.1.2   Hydrolysis of Polymer-bound Methylester to Yield Acid-functionalized Polymeric Support 9

The resin is suspended in dioxane* (20 mL/g resin) and after 15 min 1% aque-ous lithium hydroxide solution* (20 mL/g resin) is added. The mixture is shaken for 18 h at room temperature, filtered, and the resin is then washed with methy-lene chloride, THF, THF/1 N HCl (1:1), THF, methanol, methylene chloride and cyclohexane (10 mL, two times each) and dried to constant weight *in vacuo*.
Task: IR measurements and assignment of characteristic bands.

13.4.1.3   Esterification of Acid-functionalized Resin with 2-Methoxy-5-nitrobenzylester 10 (Determination of Loading)

Two 50-mg quantities of acid-functionalized resin **9** (calculation of theoretical substi-tution necessary) are placed in 2×25 ml flasks and each is shaken with 10 equiv. Cs_2CO_3 and 30 equiv. 2-methoxy-5-nitrobenzylbromide in 15 mL dry DMF. After

---

* Hazardous chemical, see Appendix

shaking overnight the resin is filtered off and washed with THF/1 N HCl (1:1) (2×5 mL), THF/H$_2$O (1:1) (1×5 mL), THF (2×5 mL), ethyl acetate and methanol (1×5 mL each). After drying the resin to constant weight *in vacuo* measurement of the IR absorption is carried out.

Task:

(a) Estimation of theoretical loading L (mol/g)

$$L = \frac{L_{old}}{1 + \Delta M \cdot L_{old}} \text{ with } L_{old} = \text{initial loading of resin (mmol/g)},$$

ΔM = difference between the molecular mass of the modified polymer-bound compound and the starting polymer-bound compound.

(b) Assignment of characteristic bands of polymer-bound 2-methoxy-5-nitrobenzylesters and comparison with acid-functionalized resin.

13.4.1.4 Determination of Loading by Means of UV-spectroscopic Analysis of 2-Methoxy-5-nitrobenzyl Alcohol 12

A defined amount of polymer-bound 2-methoxy-5-nitrobenzyl ester **11** (about 50 mg) is suspended in 10 mL dioxane and shaken for 15 min. Then 5 mL of 0.5 % LiOH is added and shaken for 24 h at room temperature. The polymer is filtered off and washed with water and THF (3×5 mL each). The filtrate is adjusted to pH = 7 with diluted hydrochloric acid and the solvent evaporated *in vacuo*. The residue is re-dissolved in 20 mL methanol, sonicated for 15 min and the amount of 2-methoxy-5-nitrobenzyl alcohol **12** released is determined with UV spectroscopy (detection at 307 nm).

Task: Calculation of loading of resin 9 using Lambert–Beer's law: λ_{max} = 307 nm, ε = 10200 mol^{-1}dm^{3}cm^{-1}, A = $\varepsilon \cdot$c$\cdot$d, with A = absorption, c = concentration mol/dm^3, d = thickness of cuvette. It is necessary to consider the weight gain of the resin after esterification.

13.4.2
Project B: Synthesis of an Antibiotic

13.4.2.1 Synthesis of 4-Iodophenylhydrazine 13

The synthesis was carried out according to the description given in the "Organikum" [6]. A solution of 1.73 g (25.1 mmol) NaNO$_2$* in 10 mL water is added dropwise over a period of 60 min at a temperature of –15 °C to a suspension of 5 g (22.9 mmol) 4-iodoaniline* in 20 mL diluted hydrochloric acid (18%). The resulting yellow-brown solution is stirred vigorously for an additional 15 min and then added in 5-mL portions to a solution of 15.5 g (68.5 mmol) SnCl$_2$·H$_2$O* in 20 mL concentrated hydrochloric acid at 10 °C. To achieve complete crystallization of the precipitate formed, the reaction mixture is cooled in an ice-bath for 1 h. The off-

---

* Hazardous chemical, see Appendix

Substance name	Symbol	Remarks
4-(methylamino)-phenol sulfate	Xn	If swallowed, seek medical advice immediately and show this container or label. Avoid release to the environment. Refer to special instructions/safety data sheets.
N-methylmorpholine	F, C	In case of accident or if you feel unwell, seek medical advice immediately (show the label if possible). Keep away from sources of ignition.
morpholine	C	In case of accident or if you feel unwell, seek medical advice immediately (show the label if possible).
palladium(II)acetate	Xi	In case of contact with eyes, rinse immediately with plenty of water and seek medical advice.
phosphoric acid	C	In case of contact with eyes, rinse immediately with plenty of water and seek medical advice. In case of accident or if you feel unwell, seek medical advice immediately (show the label if possible).
piperidine	T, F	Keep away from sources of ignition. In case of contact with eyes, rinse immediately with plenty of water and seek medical advice. In case of accident or if you feel unwell, seek medical advice immediately (show the label if possible).
potassium carbonate	Xn	After contact with skin, wash immediately with plenty of water.
2-propanol	F, Xi	Keep away from sources of ignition. In case of contact with eyes, rinse immediately with plenty of water and seek medical advice.
PyBOP	Xi	
pyridine	F, Xn	In case of contact with eyes, rinse immediately with plenty of water and seek medical advice. After contact with skin, wash immediately with plenty of water.
pyrophosphoric acid	C	In case of contact with eyes, rinse immediately with plenty of water and seek medical advice. In case of accident or if you feel unwell, seek medical advice immediately (show the label if possible).
sinapinic acid	Xi	In case of contact with eyes, rinse immediately with plenty of water and seek medical advice.
sodium azide	T+, N	In case of accident or if you feel unwell, seek medical advice immediately (show the label if possible).
sodium carbonate anhydrous	Xi	In case of contact with eyes, rinse immediately with plenty of water and seek medical advice.
sodium dithionite	Xn	In case of contact with eyes, rinse immediately with plenty of water and seek medical advice. Keep container tightly closed and dry. After contact with skin, wash immediately with plenty of water. For extinguishing, use sand, earth, powder or foam. Do not use water. Contact with acids liberates toxic gas.
sodium dodecyl sulfate	Xn	In case of contact with eyes, rinse immediately with plenty of water and seek medical advice.
sodium hydrogen sulfite	Xn	If swallowed, seek medical advice immediately and show this container or label.

Substance name	Symbol	Remarks
sodium hydroxide	C	In case of accident or if you feel unwell, seek medical advice immediately (show the label if possible). In case of contact with eyes, rinse immediately with plenty of water and seek medical advice.
sodium nitrite	O, T, N	In case of accident or if you feel unwell, seek medical advice immediately (show the label if possible). Avoid release to the environment. Refer to special instructions/safety data sheets.
sulfuric acid	C	In case of contact with eyes, rinse immediately with plenty of water and seek medical advice. Never add water to this product. In case of accident or if you feel unwell, seek medical advice immediately (show the label if possible).
tetrahydrofuran	F, Xi	Keep away from sources of ignition. Take precautionary measures against static discharges.
tetramethylethylenediamine (TEMED)	F, C	In case of accident or if you feel unwell, seek medical advice immediately (show the label if possible). In case of contact with eyes, rinse immediately with plenty of water and seek medical advice. Keep away from sources of ignition.
tetrazole	E	Keep away from heat. Keep away from sources of ignition. This material and its container must be disposed of in a safe way.
tin(II)chloride	Xn	In case of contact with eyes, rinse immediately with plenty of water and seek medical advice. In case of accident or if you feel unwell, seek medical advice immediately (show the label if possible).
triethyl amine	F, C	Keep in a cool place. Keep away from sources of ignition. In case of contact with eyes, rinse immediately with plenty of water and seek medical advice. In case of accident or if you feel unwell, seek medical advice immediately (show the label if possible).
trifluoro acetic acid	C	Keep container in a well-ventilated place. In case of contact with eyes, rinse immediately with plenty of water and seek medical advice. After contact with skin, wash immediately with plenty of water. In case of accident or if you feel unwell, seek medical advice immediately (show the label if possible). Avoid release to the environment. Refer to special instructions/safety data sheets.
triisopropylsilane	Xi	Keep away from sources of ignition. In case of contact with eyes, rinse immediately with plenty of water and seek medical advice.
Tris-buffer	Xi	In case of contact with eyes, rinse immediately with plenty of water and seek medical advice.
Triton-X114	Xn	In case of contact with eyes, rinse immediately with plenty of water and seek medical advice.

Substance name	Symbol	Remarks
zinc chloride dihydrate	C, N	Keep container tightly closed and dry. In case of accident or if you feel unwell, seek medical advice immediately (show the label if possible). Avoid release to the environment. Refer to special instructions/safety data sheets.

Subject Index

a

ab initio calculations 80
absorption of drugs 88
ACE inhibitors 62
N-acetylgalactosamine (GalNAc) 16, 18, 161, 164, 168
N-acetylglucosamine 137
acetyl protecting group 26
activation of the phosphoamidite 29
activation domain 13
active site 19, 81
adenosine 25
adenosine triphosphate 19
α-addition 173
ADME model 88
AdoMet 15
affinity 81
affinity chromatography 12, 162
affinity column 11
agarose 153
agglutinins 162
AIDS 163
alkaline phosphatase 49
alkylation of proteins 155
alkylpyrophosphates 118
allele-specific inhibition 19
aminoethylglycine 36
ammonium persulfate 154
amylose 137, 140, 143
angles of rotation 83
antibiotic 181, 186
antigen presenting cell (APC) 12
APC 12
ATP 19
AttoPhos 57
N-azidoacetylgalactosamine (GalNAz) 19
azide 17 f.

b

bacteria 13
base mismatch 31, 38
base-flipping 15
basic hydrolysis 26
BCA assay 157
benzhydrylcarbonyl-(Bhoc-) protecting group 38
benzoyl protecting group 26
binary representation 174
binding energy 82
biocatalysis 5
biochemistry, definition of 2
bioisostery 89
biological activity 171
biological effect 89
biological phenomenon 3
biological problem 5
biological target 11
biomacromolecules 3
bioorganic chemistry, definition of 2
biophase 88
biophysics 5
biosynthesis of farnesyl pyrophosphate 112
biotin 47, 161
biotin-galactose (BG) conjugate 161, 168
biotin-streptavidin complex 45, 165
biotinylated D-galactose (tracer) 165
biotransformation 88
blood coagulation cascade 171
Boc strategy 65
bovine serum albumin (BSA) 165, 167
Bradford assay 120, 156
bradykinin 61 ff.
brefeldin 19
Brookhaven Protein Data Bank 77, 92
BSA 165, 167
build-up approach 81

Chemical Biology: A Practical Course.
Edited by Herbert Waldmann and Petra Janning
Copyright © 2004 WILEY-VCH Verlag GmbH & Co. KGaA, Weinheim
ISBN: 3-527-30778-8